1st Meeting on Non-Regular Spacetime Geometry 2017

Journal of Physics: Conference Series Volume 968

Florence, Italy
20 - 22 June 2017

Editors:

Piotr T. Chrusciel
James Grant
Michael Kunzinger
Ettore Minguzzi

ISBN: 978-1-5108-5897-8
ISSN: 1742-6588

Printed from e-media with permission by:

Curran Associates, Inc.
57 Morehouse Lane
Red Hook, NY 12571

Some format issues inherent in the e-media version may also appear in this print version.

Printed by Curran Associates, Inc. (2018)

For permission requests, please contact the Institute of Physics
at the address below.

Institute of Physics
Dirac House, Temple Back
Bristol BS1 6BE UK

Phone: 44 1 17 929 7481
Fax: 44 1 17 920 0979

techtracking@iop.org

Additional copies of this publication are available from:

Curran Associates, Inc.
57 Morehouse Lane
Red Hook, NY 12571 USA
Phone: 845-758-0400
Fax: 845-758-2633
Email: curran@proceedings.com
Web: www.proceedings.com

Table of contents

Volume 968

Non-Regular Spacetime Geometry
20–22 June 2017, Florence, Italy

Accepted papers received: 1 February 2018
Published online: 22 February 2018

Preface

Non-Regular Spacetime Geometry

Peer review statement

Papers

Smoothing causal functions

Patrick Bernard and Stefan Suhr.....1

Spacetimes in Noncommutative Geometry: a definition and some examples

Fabien Besnard.....8

The annoying null boundaries

Piotr T. Chruściel and Paul Klinger.....18

Causal Fermion Systems: A Primer for Lorentzian Geometers

Felix Finster.....33

The Lorentzian distance formula in noncommutative geometry

Nicolas Franco.....47

Explicit triangular decoupling of the separated vector wave equation on Schwarzschild into scalar Regge-Wheeler equations

Igor Khavkine.....57

Connes distance and optimal transport

Pierre Martinetti.....77

Time functions and *K*-causality between measures

Tomasz Miller.....84

The representation of spacetime through steep time functions

Ettore Minguzzi.....94

On geodesics in low regularity

Clemens Sämann and Roland Steinbauer.....105

Green operators for low regularity spacetimes

Yafet Sanchez Sanchez and James Vickers.....119

On the proof of the C^0-inextendibility of the Schwarzschild spacetime

Jan Sbierski.....132

IOP Conf. Series: Journal of Physics: Conf. Series **968** (2018) 011001 doi:10.1088/1742-6596/968/1/011001

Preface

Until recently, mathematical studies of Lorentzian causality theory assumed, explicitly or implicitly, smoothness of the metric. This changed a few years ago, after independent work by Fathi and Siconolfi, and by Chruściel and Grant. Since then many papers began to explore causality under weaker conditions such as $C^{1,1}$ metrics, Lipschitz metrics, continuous metrics (and cone structures), or even topological settings. Indeed, new insights are needed when the metric is not twice-differentiable. Hence, manifolds for which the metric is not C^2 might be called non-regular, though closer inspection reveals that there are not so many differences between the C^2 and the $C^{1,1}$ theories, so one can also refer to the non-regular theories as those whose metric has regularity below $C^{1,1}$.

In Riemannian signature geometers studied non-regular manifolds using comparison geometry and other approaches. Many of these techniques have yet to be exported to the Lorentzian signature since, of course, a number of difficulties have to be addressed. In spite of this, mathematical physicists study non-regular spaces with a strong motivation that stems from the desire of understanding the ultimate nature of spacetime. The idea is that at the quantum level and hence whenever the physical conditions become extreme (say gravitational collapse, or origin of the universe) the spacetime manifold should perhaps not be approximated by a smooth manifold. If this expectation is correct then an understanding of the role of differentiability conditions in general relativity might give hints to the very nature of gravity and spacetime at the quantum scale.

The Florence meeting provided an opportunity for the mathematical relativists interested in non-regular spacetime geometry to get together with mathematical and theoretical physicists interested in non-commutative geometry, in particular in Connes' program for the unification of the fundamental forces. The meeting was also open to related contributions which could help shed light on the geometry of spacetime at the deepest (non-regular) level.

The talks were roughly divided into five interrelated subjects:

(i) Lorentzian geometry and field theory under low differentiability assumptions, (Galloway, Graf, Harris, Khavkine, Klinger, Sbierski, Steinbauer, Vickers);

(ii) Cone structures and time functions under low differentiability assumptions (Bernard, Minguzzi, Siconolfi, Suhr);

(iii) Measures and optimal transport over Lorentzian manifolds (Finster, Martinetti, Miller);

(iv) Length spaces and Alexandrov geometry (Sämann, Alexander);

(v) Non-commutative geometry and other algebraic or quantum related approaches to spacetime (Besnard, Finster, Franco, Minguzzi).

The meeting confirmed that interest in the non-regular aspects of spacetime geometry is gaining momentum. Recent publications, often by young researchers, are improving our understanding of this topic, and much is expected and hoped for in the future.

It was a pleasure to organize the first dedicated meeting on non-regular spacetime geometry, and we are thankful to the many researchers who have contributed to make it a successful event. We are happy that many of the participants contributed to this proceedings volume.

IOP Conf. Series: Journal of Physics: Conf. Series **968** (2018) 011001 doi:10.1088/1742-6596/968/1/011001

We cannot end this preface without thanking the Rector of the University of Florence, Luigi Dei, the director of the Department of Mathematics and Informatics "U. Dini" in Florence, Giorgio Ottaviani, and Professor Patrizia Pera for contributing to the financing of the meeting. Several participants benefitted from the financial support from the Department of Mathematics of Vienna University and the Austrian Science Fund FWF.

The meeting took place from June 20 to June 22, 2017, at Centro Didattico Capponi, via Gino Capponi 9, right in the city center of Florence. We thank the President of the School of Educational Sciences Gianfranco Bandini for letting us use the classrooms in such a central and wonderful location.

Piotr T. Chruściel
James Grant
Michael Kunzinger
Ettore Minguzzi

Editors/Scientific Committee
Piotr T. Chruściel
James Grant
Michael Kunzinger
Ettore Minguzzi

Local Organizing Committee
Daniel Canarutto
Ettore Minguzzi (Chair)
Omar Morandi

Participants
Stephanie Alexander, Patrick Bernard, Fabien Besnard, Piotr T. Chruściel, Felix Finster, Nicolas Franco, Greg Galloway, Melanie Graf, James Grant, Stacey Harris, Igor Khavkine, Paul Klinger, Michael Kunzinger, Pierre Martinetti, Tomasz Miller, Ettore Minguzzi, Valter Moretti, Clemens Sämann, Jan Sbierski, Antonio Siconolfi, Roland Steinbauer, Stefan Suhr, James Vickers.

IOP Conf. Series: Journal of Physics: Conf. Series **968** (2018) 011001 doi:10.1088/1742-6596/968/1/011001

A picture of the last day of the meeting. From left to right. Sämann, Harris, Minguzzi, Vickers (behind), Siconolfi, Bernard, Martinetti, Klinger (behind), Alexander, Galloway, Moretti, Finster, Besnard, Suhr, Khavkine, Miller, Sbierski, Franco, Graf, Steinbauer, Kunzinger, Grant.

Unfortunately, Piotr T. Chrusciel is missing since he had to leave on the day before.

IOP Conf. Series: Journal of Physics: Conf. Series **968** (2018) 011002 doi:10.1088/1742-6596/968/1/011002

Peer review statement

All papers published in this volume of *Journal of Physics: Conference Series* have been peer reviewed through processes administered by the proceedings Editors. Reviews were conducted by expert referees to the professional and scientific standards expected of a proceedings journal published by IOP Publishing.

IOP Conf. Series: Journal of Physics: Conf. Series **968** (2018) 012001 doi:10.1088/1742-6596/968/1/012001

Smoothing causal functions

Patrick Bernard[1] **and Stefan Suhr**[2]

[1] PSL Research University, École Normale Supérieure, and Université Paris-Dauphine, CNRS, DMA (UMR 8553), 45 rue d'Ulm, 75230 Paris Cedex 05, France

[2] Fakultät für Mathematik, Ruhr-Universität Bochum, Universitätsstraße 150, 44801 Bochum, Germany

E-mail: `patrick.bernard@ens.fr`, `stefan.suhr@rub.de`

Abstract. We describe, in the general setting of closed cone fields, the set of causal functions which can be approximated by smooth Lyapunov. We derive several consequences on causality theory.

Dans le contexte général des champs de cones fermés, on décrit l'ensemble des fonctions causales qui peuvent être approchées par des fonctions de Lyapunov lisses. On en déduit quelques conséquence en théorie de la causalité.

1. Definitions and Results

In the setting of Lorentzian geometry, the relation between continuous time functions and smooth time functions, or even better temporal functions, was not well understood before the paper of Bernal and Sanchez [3]. In the present paper, we study the more general question of the approximation of continuous causal functions by smooth ones, in the very general setting of closed cone fields, and without any causality assumption. We introduce a class of causal functions, called strictly causal functions, which can be approximated by smooth causal, and even smooth Lyapunov, functions. Examples show that some causal functions do not have this property. Yet time functions are strictly causal hence can be approximated by temporal functions, which generalizes the result of Bernal and Sanchez to the context of closed cone fields.

These considerations also have several consequences in causality theory. We recover in particular the equivalence between stable causality, the existence of a temporal function, and $\mathcal{K}$-causality.

Although it may seem less relevant for physical applications to study noncausal cone fields, we point out that our result also applies to the study of dynamical systems, where the cone field is reduced to a half-line. The relation between recurrence and Lyapunov or causal functions, respectively, for such systems was first studied by Conley. To our knowledge, our results are also new in this case.

We work on a Riemannian manifold M. A *convex cone* of the vector space E is a convex subset $C \subset E$ such that $tx \in C$ for each $t > 0$ and $x \in C$. The convex cone C is called *regular* if it is not empty and contained in an open half-space, or equivalently if there exists a linear form p on E such that $p \cdot v > 0$ for each $v \in C$. The full cone $C = E$ will be called the *singular* cone.

Definition 1. *We say that $C \subset E$ is a closed cone if it is a convex cone which is either singular or regular and if $C \cup \{0\}$ is a closed subset of E.*

IOP Conf. Series: Journal of Physics: Conf. Series **968** (2018) 012001 doi:10.1088/1742-6596/968/1/012001

Note that the empty set is a closed cone. We call it *degenerate.*

A *cone field* $\mathcal{C}$ on the manifold M is a subset of the tangent bundle TM such that $\mathcal{C}(x) := T_xM \cap \mathcal{C}$ is a convex cone for each x.

Definition 2. *We say that $\mathcal{C} \subset TM$ is a closed cone field if it is a cone field such that the union of $\mathcal{C}$ and the image of the zero section in TM is a closed subset of TM and such that $\mathcal{C}(x)$ is a closed cone for each x.*

A point $x \in M$ is called nondegenerate for $\mathcal{C}$ if $\mathcal{C}(x) \neq \emptyset$. The *domain* of $\mathcal{C}$ is the set of nondegenerate points. It is denoted by $\mathcal{D}(\mathcal{C})$. The domain of a closed cone field is closed. A cone field is called *nondegenerate* if all points are nondegenerate, *i.e.* if $\mathcal{D}(\mathcal{C}) = M$. The set of singular points (the points x such that $\mathcal{C}(x) = T_xM$) of a closed cone field is closed.

For a closed cone field $\mathcal{C}$, recall from [4] that the curve $\gamma : I \longrightarrow M$ is $\mathcal{C}$-causal (or just causal) if it is locally Lipschitz and if the inclusion $\dot\gamma(t) \in \mathcal{C}(\gamma(t)) \cup \{0\}$ holds for almost all $t \in I$. The strictly causal future $\mathcal{J}^{++}_{\mathcal{C}}(x)$ of x is the set of points $y \in M$ such that there exists a nonconstant causal curve $\gamma : [0, T] \longrightarrow M$ satisfying $\gamma(0) = x$ and $\gamma(T) = y$. The causal future $\mathcal{J}^{+}_{\mathcal{C}}(x)$ of x is the set $\mathcal{J}^{+}_{\mathcal{C}}(x) = \mathcal{J}^{++}_{\mathcal{C}}(x) \cup \{x\}$. The causal past $\mathcal{J}^{-}_{\mathcal{C}}(x)$ is the set of points $x' \in M$ such that $x \in \mathcal{J}^{+}_{\mathcal{C}}(x')$. More generally, for each subset $A \subset M$, we denote by $\mathcal{J}^{\pm}_{\mathcal{C}}(A) := \cup_{x\in A}\mathcal{J}^{\pm}_{\mathcal{C}}(x)$ the causal future and past of A. We have the inclusion $\mathcal{J}^{+}_{\mathcal{C}}(y) \subset \mathcal{J}^{+}_{\mathcal{C}}(x)$ if $y \in \mathcal{J}^{+}_{\mathcal{C}}(x)$.

Definition 3. *The function $f\colon M \to \mathbb{R}$ is said to be causal if it is continuous, and has the property that $f(y) \geqslant f(x)$ for each $y \in \mathcal{J}^{+}_{\mathcal{C}}(x)$. We denote by $\mathbf{C}$ the set of causal functions.*

Given a causal function f, the point x is said to be *neutral* if it is a singular point of $\mathcal{C}$ or if there exists $y \in \mathcal{J}^{+}_{\mathcal{C}}(x) - \{x\}$ such that $f(y) = f(x)$. Otherwise, x is said to be *strict.*

We say that $a \in \mathbb{R}$ is a *neutral value* of f if there exists a neutral point x such that $f(x) = a$. Otherwise, a is called a *strict value.*

We say that f is a *special causal function* if it is causal, and if the set of strict values of f is dense in $\mathbb{R}$. We denote by $\mathbf{S}$ the set of special causal functions. It is classical in the Conley theory of dynamical systems to consider this kind of hypothesis. The terminology *neutral values* is taken from [6].

We say that τ is a Lyapunov function if it is smooth and satisfies $d\tau_x \cdot v > 0$ for each regular point x of τ and each $v \in \mathcal{C}(x)$. We denote by $\mathbf{L}$ the set of Lyapunov functions. We have

$$\mathbf{L} \subset \mathbf{S} \subset \mathbf{C}.$$

The first inclusion following from:

Lemma 4. *Each Lyapunov function τ is special causal. Moreover, the set of neutral points of τ is contained in the set of critical points of τ.*

PROOF. If τ is a Lyapunov function and γ is a causal curve, then the function $\tau \circ \gamma$ is locally Lipschitz and satisfies $(\tau \circ \gamma)'(t) \geqslant 0$ almost everywhere, hence it is nondecreasing. This implies that the function τ is causal. If there exists $t > 0$ and a nonconstant causal curve $\gamma\colon [0, t] \to M$ such that $\tau \circ \gamma(t) = \tau \circ \gamma(0)$, then $\tau \circ \gamma$ is constant on $[0, t]$, which implies that $\gamma(0)$ is a critical point of τ. We deduce that the neutral points of τ are necessarily critical points, hence that neutral values are critical values. By Sard's Theorem, the set of regular values of τ is dense, which implies that the set of strict values is dense. □

We define the stably recurrent set $\mathcal{R}$ of $\mathcal{C}$ as the intersection of the sets of critical points of Lyapunov functions. This set is studied in [4]. In the present setting, it is proved there that:

Proposition 5. *If $x \in \mathcal{R}$, then either x is a singular point of $\mathcal{C}$, or there exists $y \in \mathcal{J}^{+}_{\mathcal{C}}(x) - \{x\}$ such that $\tau(y) = \tau(x)$ for each $\tau \in \mathbf{L}$. As a consequence*

$$\mathcal{R} = \bigcap_{\tau\in\mathbf{L}} \mathrm{neut}(\tau) = \bigcap_{\tau\in\mathbf{L}} \mathrm{crit}(\tau).$$

IOP Conf. Series: Journal of Physics: Conf. Series **968** (2018) 012001 doi:10.1088/1742-6596/968/1/012001

Moreover, there exists a Lyapunov function τ_0 such that

$$\mathcal{R} = \mathrm{neut}(\tau_0) = \mathrm{crit}(\tau_0).$$

We now state our main result, which will be proved in Section 2:

Theorem 1. *For each function $f \in \mathbf{S}$ and each $\epsilon > 0$, there exists a Lyapunov function $\tau \in \mathbf{L}$ such that $|\tau - f|(x) < \epsilon$ for each x, and which is regular outside of $\mathcal{R}$.*

To our knowledge, this statement is new even in the classical Lorentzian setting and in the setting of vector fields. See however [5] for results on smoothing causal functions.

The following example, taken from [9, Example 5.1], illustrates that the set $\mathbf{S}$ could not be replaced by $\mathbf{C}$ in the above statement.

Example 6. *We consider the manifold $M = \mathbb{S}^1 \times \mathbb{R}$, and the constant cone field $\mathcal{C}(\theta, x) = [0,\infty) \times [0,\infty) - \{(0,0)\}$. The function $f(\theta, x)$ on M is causal if and only if it is of the form $f(\theta, x) = g(x)$ for some nondecreasing function $g : \mathbb{R} \longrightarrow \mathbb{R}$. For such a causal function, all the values in the interval $g(\mathbb{R})$ are neutral. As a consequence, the function f is special causal if and only if it is constant.*

Obviously, $\mathbf{S}$, and hence $\mathbf{L}$, is not dense in $\mathbf{C}$ in this example.

Corollary 7. $\mathcal{R} = \bigcap_{f \in \mathbf{S}} \mathrm{neut}(f)$

PROOF. Since $\mathbf{L} \subset \mathbf{S}$, we have the inclusion $\bigcap_{f \in \mathbf{S}} \mathrm{neut}(f) \subset \mathcal{R}$ by Proposition 5. Conversely, consider $x \in \mathcal{R}$. If x is a singular point of $\mathcal{C}$, then it is a neutral point for each causal function. If x is not a singular point of $\mathcal{C}$, there exists $y \in \mathcal{J}^+_{\mathcal{C}}(x) - \{x\}$ such that $\tau(y) = \tau(x)$ for each $\tau \in \mathbf{L}$. The density of $\mathbf{L}$ in $\mathbf{S}$ then implies that $f(y) = f(x)$ for each special causal function f, hence x is neutral for f. □

In general the set of neutral points of a causal function can be a strict subset of $\mathcal{R}$. We recall some definitions before presenting an example. We say that f is a *time function* if it is a causal function without neutral points. Obviously, time functions are special causal. Note that the existence of a time function with our definitions prevents the presence of singularities of $\mathcal{C}$. We say that τ is a *temporal function* if it is a Lyapunov function without critical points.

Example 8. *Consider $M = S^1 \times \mathbb{R}^2$ with coordinates (θ, x, z). Equip M with the cone field*

$$\mathcal{C}(\theta, x, z) := \{(v_\theta, v_x, v_z) : v_x \geqslant 0 \text{ and } v_\theta v_x \geqslant z^2 v_\theta^2 + v_z^2\}.$$

We claim that

$$\bigcap_{f \in \mathbf{C}} \mathrm{neut}(f) = S^1 \times \mathbb{R} \times \{0\} \neq \mathcal{R} = M.$$

PROOF. Under the embedding $(\theta, x) \mapsto (\theta, x, 0)$ of $S^1 \times \mathbb{R}$ into M the cone field $\mathcal{C}$ pulls back to the cone field considered in Example 6. The coordinate function $x \colon M \to \mathbb{R}$ is a causal function of $(M, \mathcal{C})$, but not a temporal one. As seen in Example 6 the set of neutral points of every causal function includes $S^1 \times \mathbb{R} \times \{0\}$. Further the neutral points of the function x are precisely $S^1 \times \mathbb{R} \times \{0\}$. Therefore we have

$$\bigcap_{f \in \mathbf{C}} \mathrm{neut}(f) = S^1 \times \mathbb{R} \times \{0\} \subset \mathcal{R}.$$

In particular, all Lyapunov functions are critical, hence constant, on $S^1 \times \mathbb{R} \times \{0\}$. The curves $(\theta(t), x(t), z(t)) = (\theta_0 + t, x_0 + 2t + (t_0 + t)^3 - t_0^3, t_0 + t)$ and $(\theta(t), x(t), z(t)) =$

IOP Conf. Series: Journal of Physics: Conf. Series **968** (2018) 012001 doi:10.1088/1742-6596/968/1/012001

$(\theta_0 + t, x_0 + 2t + (t - t_0)^3 + t_0^3, t_0 - t)$ are causal. This implies that for every $p \in M$ we have

$$\mathcal{J}_{\mathcal{C}}^{\pm}(p) \cap (S^1 \times \mathbb{R} \times \{0\}) \neq \emptyset.$$

As a consequence, all Lyapunov functions are constant on M, hence $\mathcal{R} = M$. □

We recall that a time function is said to be Cauchy if it has the property that the image of $f \circ \gamma$ is the whole real line for each inextendible causal curve γ.

Corollary 9. *There exists a time function if and only if the stably recurrent set is empty. In this case, each causal function f can be uniformly approximated by temporal functions τ.*

If f is a Cauchy time function, then so is the temporal function τ (in this case, $\mathcal{C}$ is globally hyperbolic in the sense of [4]).

PROOF. The first claim follows directly from the definition and Proposition 5.

Let us next consider a causal function f. Since $\mathcal{R}$ is empty, there exists a time function g'. The function $g := \frac{2\epsilon}{\pi} \arctan \circ g'$ is a time function taking values in $]-\epsilon, \epsilon[$. The function $f + g$ is then a time function, hence it belongs to $\mathbf{S}$. By Theorem 1, there exists a temporal function τ such that $|\tau - (f + g)| < \epsilon$, hence $|\tau - f| < 2\epsilon$.

If f and g are two time functions such that $f - g$ is bounded, and if f is Cauchy, then so is g. Indeed, for each inextendible causal curve γ, we have $\sup(f \circ \gamma) = +\infty$, hence $\sup(g \circ \gamma) = +\infty$ and $\inf(f \circ \gamma) = -\infty$, hence $\inf(g \circ \gamma) = \infty$. □

Let us discuss the implications of our result in terms of causality relations. The stable (also called Seifert) relation $\mathcal{F}^+(x)$ is defined by

$$\mathcal{F}^+(x) := \{y : \tau(y) \geqslant \tau(x)\ \forall \tau \in \mathbf{L}\}$$

This relation can also be defined in terms of cone enlargements, see [4] for the present generality. It is an obvious consequence of Theorem 1 that

$$\mathcal{F}^+(x) = \{y : \tau(y) \geqslant \tau(x)\ \forall \tau \in \mathbf{S}\}.$$

On the other hand, it is a general fact that

$$\mathcal{K}^+(x) = \{y : \tau(y) \geqslant \tau(x)\ \forall \tau \in \mathbf{C}\} \subset \mathcal{F}^+(x)$$

is the smallest transitive closed relation containing $\mathcal{J}_{\mathcal{C}}^+$. This follows from the Auslander-Levin Theorem, [2, 8], [1, Theorem 2], as was noticed in [12, Theorem 3].

In Example 6 we have $\mathcal{K}^+(\theta, x) = \mathcal{J}_{\mathcal{C}}^+(\theta, x) = \mathbb{S}^1 \times [x, \infty)$ and $\mathcal{F}^+(\theta, x) = M$ for all $(\theta, x) \in M$. This shows that equality does not hold in general between $\mathcal{K}^+$ and $\mathcal{F}^+$. In the case where $\mathcal{R}$ is empty, however, we have by Corollary 9:

Proposition 10. *If $\mathcal{R}$ is empty, $\mathcal{K} = \mathcal{F}$.*

The equality $\mathcal{K} = \mathcal{F}$ is proved in [11] in the classical Lorentzian setting, and very recently [14, Theorem 85] under assumptions slightly stronger than the present ones.

Let us finally recover and slightly generalize some results of Minguzzi on the equivalence between $\mathcal{K}$-causality and stable causality. We say that the cone field is *$\mathcal{K}$-causal* if $\mathcal{K}^+(x) \cap \mathcal{K}^-(x) = \{x\}$ for each x. $\mathcal{K}$-causality implies that for all x and all $y \in \mathcal{J}_{\mathcal{C}}^+(x) - \{x\}$ there exists $f \in \mathbf{C}$ with $f(y) > f(x)$. Recall that the cone field is *stably causal* if $\mathcal{R}$ is empty. This is equivalent to *$\mathcal{F}$-causality*, that is to the property that $\mathcal{F}^+(x) \cap \mathcal{F}^-(x) = \{x\}$ for each x, as follows from Proposition 5, see also [4, 7, 10, 16].

Proposition 11. *The following properties are equivalent:*

IOP Conf. Series: Journal of Physics: Conf. Series **968** (2018) 012001 doi:10.1088/1742-6596/968/1/012001

- *The cone field $\mathcal{C}$ is stably causal.*
- *There exists a temporal function.*
- *There exists a time function.*
- *The cone field is $\mathcal{K}$-causal.*

PROOF. We have already proved the equivalence of the three first points.

Since $\mathcal{K}^+(x) \cap \mathcal{K}^-(x) \subset \mathcal{F}^+(x) \cap \mathcal{F}^-(x)$, any of the first three conditions implies the last point.

Conversely, let us consider the set $\mathbf{C}([\mathbf{0}, \mathbf{1}])$ of causal functions taking values in $[0, 1]$, endowed with the topology of uniform convergence on compact sets. It is a separable metric space, as a subset of the separable metric space of all continuous functions with the topology of uniform convergence on compact sets. We consider a dense sequence $f_i \in \mathbf{C}([\mathbf{0}, \mathbf{1}])$ and $f := \sum_{i \geqslant 1} 2^{-i} f_i$, which is an element of $\mathbf{C}$. If the cone field is $\mathcal{K}$-causal, then f is a time function. Indeed, if $y \in \mathcal{K}^+(x) - \{x\}$, then $f_i(y) \geqslant f_i(x)$ for each i, and there exists j such that $f_j(y) > f_j(x)$. As a consequence, we get that $f(y) > f(x)$. □

2. Proof of Theorem 1

We need more definitions from [4].

We say that $\mathcal{E} \subset TM$ is an open cone field if it is a cone field which is open as a subset of TM. Then $\mathcal{E}(x)$ is an open cone for each x. We say that $\mathcal{E}$ is an open enlargement of $\mathcal{C}$ if it is an open cone field containing $\mathcal{C}$.

Given an open cone field $\mathcal{E}$, we say that the curve $\gamma : I \longrightarrow M$ is $\mathcal{E}$-timelike (or just timelike) if it is piecewise smooth and if $\dot{\gamma}(t) \in \mathcal{E}(\gamma(t))$ for all t in I (at nonsmooth points, the inclusion is required to hold for left and right derivatives). The *chronological future* $\mathcal{I}^+_{\mathcal{E}}(x)$ of x is the set of points $y \in M$ such that there exists a nonconstant timelike curve $\gamma : [0, T] \longrightarrow M$ satisfying $\gamma(0) = x$ and $\gamma(T) = y$. The *chronological past* $\mathcal{I}^-_{\mathcal{E}}(x)$ of x is the set of points $x' \in M$ such that $x \in \mathcal{I}^+_{\mathcal{E}}(x')$. More generally, for each subset $A \subset M$, we denote by $\mathcal{I}^\pm_{\mathcal{E}}(A) := \cup_{x \in A} \mathcal{I}^\pm_{\mathcal{E}}(x)$ the *chronological future and past of* A. The chronological future can also be defined using only smooth timelike curves, see [4, Lemma 22].

Proposition 12. *Let f be a special causal function, and let $a \in \mathbb{R}$ a strict value of f. Then for each open interval $]a_-, a_+[\ni a$, there exists a Lyapunov function $\tau : M \longrightarrow [a_-, a_+]$ such that $\tau(x) = a_+$ if $f(x) \geqslant a_+$ and $\tau(x) = a_-$ if $f(x) \leqslant a_-$ and all values in $]a_-, a_+[$ are regular for τ.*

PROOF OF THEOREM 1 FROM PROPOSITION 12: We fix a special causal function f and $\epsilon > 0$. For each k, there exists a Lyapunov function $\tau_k : M \longrightarrow [k\epsilon, (k+1)\epsilon]$ such that $\tau_k = (k+1)\epsilon$ on $\{f \geqslant (k+1)\epsilon\}$ and $\tau_k = k\epsilon$ on $\{f \leqslant k\epsilon\}$. The function τ which is equal to τ_k on the set $f^{-1}([k\epsilon, (k+1)\epsilon])$ for each k is a Lyapunov function, whose set of critical values is $\epsilon\mathbb{Z}$, and which satisfies $|f - \tau| < \epsilon$.

In order to control the critical set, we consider a Lyapunov function $\tau_0 : M \longrightarrow [0, \epsilon]$ which is regular outside of $\mathcal{R}$. Then $\tau + \tau_0$ is regular outside of $\mathcal{R}$ and $|f - (\tau + \tau_0)| < 2\epsilon$. □

PROOF OF PROPOSITION 12: Let $a \in \mathbb{R}$ be a strict value of the special causal function $f \colon M \to \mathbb{R}$. Then $U := \{f > a\}$ is open with $\mathcal{J}^{++}_{\mathcal{C}}(\overline{U}) \subset U$ and ∂U consists only of regular points of $\mathcal{C}$. Especially there exists a neighborhood of ∂U consisting only of regular points of $\mathcal{C}$ since the singular set of $\mathcal{C}$ is closed. Recall from [4] the definition of a trapping domain:

Definition 13. *An open set $A \subset M$ is a trapping domain for the open cone field $\mathcal{E}$ if $\mathcal{I}^+_{\mathcal{E}}(A) \subset A$. A is a trapping domain for the closed cone field $\mathcal{C}$ if it is a trapping domain for some open enlargement $\mathcal{E}$ of $\mathcal{C}$. We say that the domain is locally trapping at a point $x \in \partial A$ if there exists an open neighborhood U of x such that $A \cap U$ is a trapping domain for $\mathcal{C}_{|U}$.*

IOP Conf. Series: Journal of Physics: Conf. Series **968** (2018) 012001 doi:10.1088/1742-6596/968/1/012001

We proved in [4, Lemma 35] that a domain is trapping if and only if it is locally trapping at each point of its boundary.

Proposition 14. *Let U be an open set such that $\mathcal{J}_{\mathcal{C}}^{++}(\bar{U}) \subset U$ and assume that ∂U consists only of regular points of $\mathcal{C}$. Further let F_i be a closed set contained in U, and let F_e be a closed set disjoint from $\bar{U}$. Then there exists a trapping domain A such that $F_i \subset A$ and F_e is disjoint from $\bar{A}$.*

PROOF OF PROPOSITION 14: Set $W := M - \bar{U}$. For an open cone field $\mathcal{E}$ denote with $\mathcal{E}_W$ the open cone field which is equal to $\mathcal{E}$ on W and empty on $\bar{U}$. We denote by $\partial^- \mathcal{I}^+_{\mathcal{E}_W}(y)$ the part of the boundary of $\mathcal{I}^+_{\mathcal{E}_W}(y)$ which is contained in W. Note that the open set $\mathcal{I}^+_{\mathcal{E}_W}(y)$ is locally trapping at each point of $\partial^- \mathcal{I}^+_{\mathcal{E}_W}(y)$ and $\mathcal{I}^+_{\mathcal{E}_W}(y) \subset W$.

Lemma 15. *For each $x \in \partial U$, there exists $y_x \in W \cap \mathcal{J}_{\mathcal{C}}^-(x)$ and an enlargement $\mathcal{E}^x$ such that $V_x := \mathcal{I}^+_{\mathcal{E}^x_W}(y_x)$ is bounded in M and disjoint from F_e.*

PROOF OF LEMMA 15: Let us fix $x \in \partial U$. We consider a bounded open neighborhood V of x in M, which is disjoint from F_e and does not contain any singular point of $\mathcal{C}$. We furthermore assume that there exists a temporal function τ on V, which satisfies $d\tau_z \cdot v > 2|v|_z$ for each $z \in V$ and $v \in \mathcal{C}(z)$. We will prove that there exists $y^x \in W \cap \mathcal{J}_{\mathcal{C}}^-(x)$ and an open enlargement $\mathcal{E}^x$ of $\mathcal{C}$ such that $\mathcal{I}^+_{\mathcal{E}^x_W}(y^x) \subset V$. Since V is bounded so is $\mathcal{I}^+_{\mathcal{E}^x_W}(y^x)$. Further $\mathcal{I}^+_{\mathcal{E}^x_W}(y^x)$ is disjoint from F_e because V is.

We consider a nested sequence of open enlargements $\mathcal{E}^n$ with $\mathcal{C} \subset \mathcal{E}^n \subset \mathcal{E}$ and $\cap_n \mathcal{E}^n = \mathcal{C}$. The existence of such a sequence is proved in [4, Lemma 20]. We set $L = d(x, \partial V)/2$ where d denotes the distance with respect to the Riemannian metric.

We first prove the existence of a $\mathcal{C}$-causal curve $\eta : [-\delta, 0]$, parametrized by arclength, such that $\eta(0) = x$ and $\eta(t) \in W$ for each $t \in [-\delta, 0[$. For this, we consider a sequence $\eta_n : [-L, 0] \longrightarrow V$ of $\mathcal{E}_n$-timelike curves parametrized by arclength. At the limit (possibly along a subsequence), we obtain a curve $\tilde{\eta} : [-L, 0] \longrightarrow V$, which is $\mathcal{C}$ causal according to the limit curve Lemma [4, Lemma 23].

Since $\cap \mathcal{E}^n = \mathcal{C}$, the curves η_n satisfy $d\tau_{\eta_n(t)} \cdot \dot{\eta}_n(t) \geqslant |\dot{\eta}_n(t)| = 1$ for each $t \in [-L, 0]$ when n is large enough, hence $\tau \circ \eta_n(-t) \leqslant \tau \circ \eta_n(0) - t$ for each $t \in [0, L]$. At the limit, we deduce that the curve $\tilde{\eta}$ is not constant. We parametrize it by arclength and get η. Since $\mathcal{J}_{\mathcal{C}}^{++}(\bar{U}) \subset U$ and $\eta(0) \notin U$, we obtain that $\eta(t) \in W$ for $t < 0$.

We now consider a sequence $y_n \in \mathcal{J}_{\mathcal{C}}^-(x) \cap W$ converging to x. These points can be taken on the curve η. We claim that $\mathcal{I}^+_{\mathcal{E}^n_W}(y_n)$ is contained in V when n is large enough. If we assume the contrary, there exists for each n large enough a smooth $\mathcal{E}^n$-timelike curve $\gamma_n : [0, L] \longrightarrow V \cap W$ parametrized by arclength and starting at y_n. As above, these curves converge uniformly to a nonconstant limit $\gamma : [0, L] \longrightarrow \overline{W}$ which satisfies $\gamma(0) = x$. The fact that $\gamma(L)$ belongs to $\overline{W} \subset M - U$ is in contradiction with the hypothesis $\mathcal{J}_{\mathcal{C}}^{++}(x) \subset U$. □

The set $A := \cup_x V_x \cup \bar{U}$ is open in M. The boundary of A is contained in $\cup_x \partial^- V_x$, hence A is locally trapping at each point of its boundary, hence it is trapping by [4, Lemma 35]. Since $\overline{U} \subset A$ we know that $F_i \subset A$. Further by construction $\overline{A}$ is disjoint from F_e. This ends the proof of Proposition 14. □

In order to finish the proof of Proposition 12 we invoke two propositions from [4]:

Proposition 16. *Let A be a trapping domain, let F_i be a closed set contained in A, let F_e be a closed set disjoint from $\bar{A}$, and let θ be a point in the boundary of A.*

Then there exists a smooth (near $\mathcal{D}(\mathcal{C})$) trapping domain A' which contains F_i, whose boundary contains θ, and whose closure is disjoint from F_e.

Proposition 17. *Let A be smooth trapping domain, then there exists a (smooth) Lyapunov function $\tau' : M \longrightarrow [-1,1]$ such that $A = \{\tau' > 0\}$ and all values in $]-1,1[$ are regular values of A (hence $\partial A = \{\tau' = 0\}$).*

If F_i and F_e are closed sets contained in A and disjoint from $\bar{A}$, respectively, we can moreover impose that $\tau' = 1$ on F_i and $\tau' = -1$ on F_e.

Here we consider $F_i := \{f \geqslant a_+\}$ and $F_e := \{f \leqslant a_-\}$. The Lyapunov function τ claimed in Proposition 12 is defined as

$$\tau := \frac{1}{2}[(a_+ - a_-)\tau' + (a_+ + a_-)].$$

This ends the proof of Proposition 12.

Acknowledgments

The research leading to these results has received funding from the European Research Council under the European Union's Seventh Framework Programme (FP/2007-2013) / ERC Grant Agreement 307062. Stefan Suhr is supported by the SFB/TRR 191 'Symplectic Structures in Geometry, Algebra and Dynamics', funded by the DFG.

References

[1] Akin E and Auslander J 2010 *Studia Math.* **201** 49–63
[2] Auslander J 1964 *Contr. Differential Equations* **3** 65–74
[3] Bernal A N and Sánchez M 2005 *Comm. Math. Phys.* **257** 43–50
[4] Bernard P and Suhr S 2017 Lyapounov Functions of closed Cone Fields: from Conley Theory to Time Functions, to appear in Comm. Math. Phys. *Preprint* arXiv:1512.08410v2
[5] Chruściel P Grant J and Minguzzi E 2016 *Ann. Henri Poincaré* **17** 2801–2824
[6] Fathi A and Pageault P 2015 *Ergod. Th. Dynam. Sys.* **35** 1187–1207
[7] Hawking S W and Sachs R K 1974 *Commun. Math. Phys.* **35** 287–296
[8] Levin V L 1983*Soviet Math. Dokl.* **28** 715–718
[9] Minguzzi E 2007 *Class. Quantum Grav.* **25** 015010
[10] Minguzzi E 2009a *J. Geom. Phys.* **59** 827–833
[11] Minguzzi E 2009b *Comm. Math. Phys.* **209** 239–248
[12] Minguzzi E 2011 *Advances in Lorentzian geometry* ed M Plaue, A Rendall, M Scherfner AMS/IP Stud. Adv. Math. vol 49 (Providence: Amer. Math. Soc.) 97–104
[13] Minguzzi E 2016 *Classical Quantum Gravity* **33** 115001
[14] Minguzzi E 2017 Causality theory for closed cone structures with applications *Preprint* arXiv:1709.06494
[15] Minguzzi E and Sánchez M 2008 *Recent developments in pseudo-Riemannian geometry* ed D. V. Alekseevsky and H. Baum (Zürich: Eur. Math. Soc.) 299–358
[16] Seifert H 1971 *Gen. Relativ. Gravit.* **1** 247–259

IOP Conf. Series: Journal of Physics: Conf. Series **968** (2018) 012002 doi:10.1088/1742-6596/968/1/012002

Spacetimes in Noncommutative Geometry: a definition and some examples

Fabien Besnard

Fabien Besnard, EPF, 3 bis rue Lakanal, Sceaux F-92330

E-mail: fabien.besnard@epf.fr

Abstract. We quickly review the notion of spectral triples - the noncommutative generalization of Riemannian spin manifolds devised by Alain Connes - and propose extensions of it to the Lorentzian and anti-Lorentzian signatures, which we call "spectral spacetimes". We motivate this definition and give several examples.

1. Introduction

There are several meanings one can give to the words "noncommutative spacetime". In this short review we will be concerned with the extension to Lorentzian (mostly plus) or anti-Lorentzian (mostly minus) signature of Alain Connes' Noncommutative Geometry. Several proposals in this direction have already been laid down ([1], [2], [3], [4]), to which we unfortunately cannot do justice here (a more thorough review can be found in [5]). Our approach differs only on a few points from previous ones but these turn out to have important consequences. In particular we can build an elementary and most natural example (section 5.3) which is not covered by existing frameworks. Another important difference is that the C^*-structure is an output rather than an input in our approach.

The paper is organized as follows: in section 2 we recall how the classical theorems of Gelfand and Naimark give a spectral characterization of (locally compact) topological spaces which allows for a noncommutative generalization. This duality is extended to spin geometry in the next section, where we review the definition of spectral triples and give a simple but important commutative and finite example: the canonical spectral triple over a finite weighted graph. In particular we explain how the geodesic distance can be recovered, thanks to a formula of Connes, both in the discrete and the manifold cases. In section 4 we explain the strategy which can be used to generalize Noncommutative Geometry to other, in particular Lorentzian, signatures. The only original proposal in this text is in section 5, where we put forward the definition of "spectral spacetimes". We give some examples of this structure, in particular a discrete example which is canonically constructed over a finite weighted directed graph and is the obvious anti-Lorentzian counterpart of the construction reviewed in section 3. We also define a noncommutative generalization of this example, which we call the "split Dirac structure".

2. The spectral characterization of space

Two classical theorems of Gelfand and Naimark are often viewed as having opened the doorway to the world of Noncommutative Geometry. In order to state them, let us recall some definitions.

IOP Conf. Series: Journal of Physics: Conf. Series **968** (2018) 012002 doi:10.1088/1742-6596/968/1/012002

Definition 1 *A* C^*-algebra *is a complete normed* $\mathbb{C}$*-algebra* $\mathcal{A}$ *with an involution* $*$ *such that for all* $a \in \mathcal{A}$, $\|a^*a\| = \|a\|^2$. *An element of* $\mathcal{A}$ *is said to be* positive *iff it is of the form* a^*a *for some* a. *A linear functional* ω *on* $\mathcal{A}$ *is said to be a* state *iff it has norm* 1 *and* $\omega(a) \in \mathbb{R}^+$ *for all positive* $a \in \mathcal{A}$. *A state is said to be* pure *if it cannot be decomposed as a non-trivial convex combination of other states.*

Let us give some examples.

(i) Let X be a locally compact Hausdorff topological space. Then the algebra $\mathcal{C}_0(X)$ of complex continuous functions on X vanishing at infinity, equipped with the supremum norm and complex conjugation of functions, is a commutative C^*-algebra. If X is compact, then $\mathcal{C}_0(X) = \mathcal{C}(X)$ and the algebra is unital. On $\mathcal{C}_0(X)$ the pure states are the evaluation maps $\hat{x} : f \mapsto f(x)$, which send a function f to its value at x.

(ii) Let $\mathcal{H}$ be a Hilbert space and $B(\mathcal{H})$ be the algebra of bounded operator on $\mathcal{H}$ equipped with the Hilbert adjoint $*$ and the operator norm. Then any subalgebra $\mathcal{A}$ of $B(\mathcal{H})$ which is stable by $*$ and norm closed is a C^*-algebra. In particular $B(\mathcal{H})$ itself is a C^*-algebra. For any vector ψ of norm 1, one can define $\omega_\psi : a \mapsto \langle\psi, a\psi\rangle$, which can be proven to be a pure state on $B(\mathcal{H})$. (These are not the only pure states on this algebra.)

The theorems proven by Gelfand and Naimark in the 1940's assert that there are essentially no other example of C^*-algebras besides the two above. More precisely, according to the first theorem, any commutative C^*-algebra $\mathcal{A}$ is canonically isomorphic to the algebra of continuous functions vanishing at infinity defined on its space of pure states $\mathcal{P}(\mathcal{A})$, which turns out to be Hausdorff and locally compact. The isomorphism is the map $a \mapsto (\omega \mapsto \omega(a))$, i.e. a is identified by biduality to the map which evaluates pure states on a. The second theorem shows that any C^*-algebra has a unitary faithful representation on some Hilbert space, i.e. is isomorphic to a C^*-algebra of the kind given above as second example.

In view of these theorems, giving a commutative C^*-subalgebra of $B(\mathcal{H})$, for some Hilbert space $\mathcal{H}$, or giving a locally compact Hausdorff space amount to the same. The interest of the first point of view is that it can be generalized merely by removing the hypothesis of commutativity, defining the field of "Noncommutative Topology" (for more on this, see [6]).

3. Spectral triples, Connes' distance formula, noncommutative 1-forms

3.1. Definition

In order to go beyond topology, one needs to enrich the structure. Alain Connes found an algebraic equivalent of Riemannian spin manifolds, in the spirit of Gelfand and Naimark.

Definition 2 *A* real even spectral triple *is a multiplet* $S = (\mathcal{A}, \mathcal{H}, \pi, D, C, \chi)$ *with* $\mathcal{A}$ *a* C^**-algebra,* $\mathcal{H}$ *a Hilbert space,* π *a representation of* $\mathcal{A}$ *on* $\mathcal{H}$*,* D *and* χ *linear operators and* C *an antilinear operator, such that:*

(i) $\chi^2 = 1$, $\chi^* = \chi$, $[\chi, \pi(\mathcal{A})] = 0$, $\{\chi, D\} = 0$,

(ii) D *is densely defined, self-adjoint, and satisfy additional analytical conditions,*

(iii) $\mathcal{A}_{\text{Lip}} := \{a \in \mathcal{A} | [D, \pi(a)]$ *is bounded* $\}$ *is dense in* $\mathcal{A}$,

(iv) $C^2 = \pm 1$, $C^*C = 1$, $\{C, D\} = 0$, $C\chi = \pm\chi C$.

In the definition above $\{.,.\}$ stands for the anti-commutator. Some comments are in order.

(i) We have only quoted the real even case, since this is the only one we will deal with. It corresponds to Riemannian spin manifolds of even dimension in the commutative case.

(ii) The operator D is called the Dirac operator. The analytical conditions which we have omitted concern the growth of its eigenvalues.

(iii) The signs in (iv) define the KO-dimension of S, which is an integer modulo 8. We refer to [7] for details.

Given a Riemannian spin manifold (M, g, σ), with g the metric and σ the spin structure, one can build its so-called *canonical triple* $S_{\text{can}}(M, g, \sigma)$ where the algebra is $\mathcal{C}_0(M)$, $\mathcal{H}$ is the Hilbert space of square integrable spinor fields, D is the usual Dirac operator, given locally by $D = i\sum_k e_k \nabla^S_{e_k}$, where (e_k) is a moving frame and ∇^S is the spin connection, χ is the chirality element of the Clifford bundle, which defines the orientation, and C is the charge conjugation operator, which selects the real form $Cl(M, g)$ of the complex Clifford bundle $\mathbb{C}l(M)$ (it commutes with real vector fields). Under some regularity conditions, $S_{\text{can}}(M, g, \sigma)$ is the only example of spectral triple with a commutative algebra (Connes' reconstruction theorem [8]), up to $D \to D + \rho$, where ρ is a linear operator which acts fibrewise on the spinor fields by $\psi \mapsto (x \mapsto \rho(x)\psi(x))$, with $\rho(x)$ an endomorphism of the spinor module at $x \in M$.

Though the proof of the reconstruction theorem is extremely involved, the metric is quite easily recovered through the geodesic distance d_g it defines. This is done thanks to the following proposition.

Proposition 1 *([9], chap. VI) Let S be a spectral triple with algebra $\mathcal{A}$, and let $\mathcal{P}(\mathcal{A})$ be the space of pure states of $\mathcal{A}$. Then the formula*

$$d_C(\omega, \omega') = \sup_{a \in \mathcal{A}_{\text{Lip}}} \{|\omega(a) - \omega'(a)|, \|[D, \pi(a)]\| \leq 1\} \tag{1}$$

defines a (possibly infinite) distance on $\mathcal{P}(\mathcal{A})$ (Connes' distance). If $S = S_{\text{can}}(M, g, \sigma)$, then $d_C = d_g$.

The crucial point in the proof of the second part of the proposition, is the observation that if a is a smooth complex function, $[D, \pi(a)] = i\gamma(da)$, where $\gamma : \Omega^1(M) \to \mathbb{C}l(M)$ is the map which identifies (complex) 1-forms on M with sections of the Clifford bundle, through the musical isomorphism and the canonical inclusion of TM into $Cl(M, g)$. As an aside, we immediately get the equality

$$\{\sum_i \pi(a_i)[D, \pi(b_i)], a_i, b_i \in \mathcal{C}^\infty(M)\} = \gamma(\Omega^1(M)) \tag{2}$$

The left-hand side of (2) does not make use of commutativity. Hence, for any spectral triple $S = (\mathcal{A}, \pi, \ldots)$, we can define noncommutative 1-forms to be the elements of $B(\mathcal{H})$ of the form $\omega = \sum_i \pi(a_i)[D, \pi(b_i)]$, where a_i, b_i are "smooth" elements in $\mathcal{A}$, that is, elements on which the derivation $[|D|, .]$ can act an arbitrary number of times.

3.2. A discrete example

One of the virtue of Noncommutative Geometry is its ability to handle the discrete and the continuous objects within the same formalism. As a very simple example, consider two points separated by a distance $\delta > 0$. Let us consider the spectral triple with algebra $\mathcal{A} = \mathbb{C}^2$, representation $\pi(a_1, a_2) = \begin{pmatrix} a_1 & 0 \\ 0 & a_2 \end{pmatrix}$ acting on $\mathcal{H} = \mathbb{C}^2$ equipped with the canonical scalar product, and Dirac operator $D = \frac{1}{\delta}\begin{pmatrix} 0 & 1 \\ 1 & 0 \end{pmatrix}$ (the chirality and charge conjugation do not matter for the moment). Since $[D, \pi(a_1, a_2)] = \frac{a_2 - a_1}{\delta}\begin{pmatrix} 0 & 1 \\ -1 & 0 \end{pmatrix}$, it is immediate to check that Connes' distance between the two points (seen as pure states) is exactly δ.

It turns out that the generalization of this example to any finite metric space is not as obvious as one could expect. Since this will be important in the sequel, let us consider in some detail the

IOP Conf. Series: Journal of Physics: Conf. Series **968** (2018) 012002 doi:10.1088/1742-6596/968/1/012002

case of a finite weighted graph $G = (V, E, \delta)$, with V the set of vertices, E the set of edges, and $\delta : E \to]0, +\infty[$ a weight function. This can be seen as a finite model of a manifold. The weight function and the adjacency structure of the graph give rise to a geodesic distance d on V. We now want to build a spectral triple such that its Connes' distance is equal to d. The solution, known for a long time ([10]), is to use a representation of the algebra with non-trivial multiplicities. We will present it here in a systematic way using the notion of *split graph* ([5]). The split graph of G is $\tilde{G} = (\tilde{V}, \tilde{E}, \tilde{\delta})$, where $\tilde{V} = E \times \{-;+\}$, $\tilde{E} = \{((e,-);(e,+))|e \in E\}$, and $\tilde{\delta}((e,-);(e,+)) = \delta(e)$. Since $\tilde{E} \simeq E$, this can be seen as a splitting of G into the disconnected union of its edges. Let us consider an orientation of the graph G, seen as a couple (s,t) of maps (source, target) from E to V. We will use the shorthand notation $e_- := s(e)$ and $e_+ := t(e)$. We can now define the so-called canonical spectral triple $S_{\text{can}}(G, s, t)$. The algebra is $\mathcal{A} = \mathcal{C}(V) \simeq \mathbb{C}^{|V|}$, the Hilbert space is $\mathcal{H} = \mathcal{C}(\tilde{V}) \simeq \mathbb{C}^{2|E|}$, equipped with the L^2 scalar product, and the representation is defined by

$$(\pi(a)F)(e,\pm) = a(e_\pm)F(e,\pm) \tag{3}$$

for every $a \in \mathcal{A}$, $F \in \mathcal{H}$ and $e \in E$. Finally the Dirac operator, chirality and real structures are given by:

$$DF(e,\pm) = \frac{1}{\delta(e)}F(e,\mp); \quad \chi F(e,\pm) = \pm F(e,\pm), \quad CF(e,\pm) = \pm\overline{F(e,\pm)} \tag{4}$$

One can check that with these definitions, Connes' distance exactly matches the geodesic distance defined by the weight δ on the graph G ([5]).

4. NCG in the anti-Lorentzian signature

4.1. Features common to all approaches

As pointed out in the introduction, many authors have investigated the question of extending Connes' Noncommutative Geometry to semi-Riemannian signature. In all approaches, the main difference with the Riemannian case is that the Dirac operator is no longer self-adjoint with respect to a scalar product, but with respect to a non-degenerate non-positive product, that in the sequel we will call *the Krein product*. The reason for this is the following: given a space and time oriented spin manifold (M, g, σ) with g of signature (p,q), and its spinor bundle S, there is a canonical family of non-degenerate products defined on each fibre S_x by the requirement that each vector $v \in T_xM$ must act in a self-adjoint way ([11]). That is, $(.,.)_x$ is a product in this family if, and only if:

$$(v \cdot \psi, \psi')_x = (\psi, v \cdot \psi')_x \tag{5}$$

for each $v \in T_xM$, $\psi, \psi' \in S_x$. It is easy to see that the different products in the family are non-zero real multiple of one another. The existence of a time orientation allows one to glue these products together in order to form a smooth non-degenerate spinor metric on S ([12]), and finally to define a non-degenerate product on $\Gamma_c(S)$, that is, on compactly supported spinor fields, through

$$(\Psi, \Psi') = \int_M (\Psi(x), \Psi'(x))_x \mathrm{vol}_g \tag{6}$$

where the integral is with respect to the canonical semi-Riemannian volume form. It is then possible to find a differential form ω such that $\eta := \gamma(\omega)$ is a fundamental symmetry, i.e. it satisfies the following requirements: $\eta^2 = \mathrm{Id}$, $\eta^\times = \eta$, where $\times$ is the adjunction of operators for the product $(.,.)$, and $(\Psi, \eta \cdot \Psi) > 0$ for all non-zero spinor field Ψ. Hence η turns the indefinite product $(.,.)$ into a scalar product $\langle .,. \rangle_\eta = (., \eta .)$. The completion $\mathcal{H}$ of $\Gamma_c(S)$ with respect to $\langle .,. \rangle_\eta$ is thus a Hilbert space, on which a non-degerate product $(.,.)$ is defined, related to the scalar product by the fundamental symmetry η. This is called a Krein space ([13]). The

Dirac operator $D = i\sum_k g(e_k, e_k)e_k\nabla^S_{e_k}$ is formally Krein-self adjoint with respect to $(.,.)$ and is even essentially Krein self-adjoint in the case where the metric can be rotated to a geodesically complete Riemannian one (see [14], Satz 3.19).

As a consequence, in the existing literature on semi-Riemannian Noncommutative Geometry, one replaces the Hilbert space with a Krein space and D is required to be Krein self-adjoint in some sense ([1], [2], [3], [4]). The fundamental symmetry is often seen to be part of the structure ([2], [3]).

4.2. The signature problem

The previous approaches, however, do not distinguish between the different signatures. It is hence impossible to speak of Lorentzian noncommutative manifold, for instance. We here put forward a proposal in this direction. The idea comes from the following theorem (here anti-Lorentzian means signature $(+,-,\dots,-)$, Lorentzian means signature $(-,+,\dots,+)$):

Theorem 1 *([12]) Let (M, g) be a semi-Riemannian space and time orientable spin manifold of even dimension. Then (M, g) is*

(i) anti-Lorentzian iff there exists a never vanishing 1-form β such that $C\beta = \beta C$ and $(.,\gamma(\beta)^{-1}.)$ is positive definite.

(ii) Lorentzian iff there exists a never vanishing 1-form β such that $C\beta = \beta C$ and $(.,\gamma(\beta)^{-1}\chi.)$ is positive definite.

Since a good theory of noncommutative 1-forms is available, this motivates the definition of spectral spacetimes, to come in the next section. Moreover, the form β above is exact iff it is the differential of a temporal function, and by a theorem of Sánchez ([15], [16]), this shows that a manifold is anti-Lorentzian and stably causal iff there exists an exact 1-form with the above property, thus providing a way to generalize stable causality in the noncommutative setting (in a way similar to [3]).

5. Spectral spacetimes

5.1. Definition

Motivated by the above considerations, we put forward the following definition.

Definition 3 *An* even anti-Lorentzian spectral spacetime *is a multiplet $S = (\mathcal{A}, K, (.,.), \pi, D, C, \chi)$, where:*

(i) $(K, (.,.))$ is a Krein space,

(ii) $\mathcal{A}$ is an algebra and π is a representation of it on K,

(iii) D is an operator on K such that $D^\times = D$,

(iv) $\chi \in B(K)$ is such that $\chi^2 = 1$, $[\pi(a), \chi] = 0$ for all $a \in \mathcal{A}$, $\chi D = -D\chi$ and $\chi^\times = -\chi$,

(v) C is an antilinear operator on K such that $C^2 = \pm 1$, $CD = -DC$, $C\chi = \pm\chi C$, $C^\times C = 1$.

(vi) there exists a noncommutative 1-form β, called a time-orientation form, *such that $[C, \beta] = 0$ and $\langle .,.\rangle_\beta := (.,\beta^{-1}.)$ is positive definite.*

Such a spectral spacetime is said to be stably causal *if β can be chosen to be of the form $\beta = [D, \pi(a)]$ for some $a \in \mathcal{A}$.*

The Lorentzian case is obtained by inserting a chirality operator in the definition of $\langle .,.\rangle_\beta$, just as in theorem 1. Let us observe that it is not straightforward to generalize this definition to other signatures, such as $(2, n-2)$, since the theory of noncommutative p-forms, for $p > 1$ involves the notion of junk, so that these forms cannot be made to act on K ([17]). In the sequel

IOP Conf. Series: Journal of Physics: Conf. Series **968** (2018) 012002 doi:10.1088/1742-6596/968/1/012002

"spectral spacetime" without any other qualification will always mean even anti-Lorentzian spectral spacetime.

An example of this structure is of course given by the *canonical spectral spacetime* $S_{\text{can}}(M, g, \sigma)$ defined out of an even dimensional space and time orientable spin anti-Lorentzian manifold (M, g, σ) in the way described in section 4.1. However, it is important to note that definition 3 is incomplete since we do not take care of the analytical conditions. In particular we cannot know whether $\mathcal{A}$ stands for an algebra of smooth or continuous functions, or some other degree of regularity. Of course, this is sufficient to treat the finite-dimensional case. Since we already know what a spacetime is in the commutative case, we can apply our definition in the almost-commutative case, which is, up to now, the only one needed for particle physics (there are two possibilities: tensorizing a Lorentzian manifold with a finite spectral triple or an anti-Lorentzian manifold with an "antieuclidean spectral triple", details are given in [5]). Moreover, in order to apply the stable causality condition as stated in the definition to the classical case, we must add a unit to the algebra, or equivalently consider a compactification of the manifold. Alternatively we can restrict to a compact subset of the original manifold.

5.2. How to recover a C^*-algebra

At first sight, definition 3 seems to be in sharp contrast with the definition of spectral triple on an essential point: no C^*-structure is postulated on $\mathcal{A}$. This may look strange, since this is the way a topological space, independent of any notion of metric, can be recovered out of a commutative algebra, as explained in section 2. In truth, C^*-structures have not disappeared. They are defined through the scalar products $\langle .,. \rangle_\beta$, where β runs over the space of time-orientation forms. Each such scalar product determines an involution $*_\beta$ on operators, hence a C^*-structure on $B(K)$. In the case $\mathcal{A} = \mathcal{C}(M)$, we have $f^{*_\beta} = \bar{f} \in \mathcal{A}$ for every β. We thus see that for all β, $*_\beta$ defines a C^*-involution on $\mathcal{C}(M)$ which is unique. In the other cases there is no guarantee that $*_\beta$ stabilizes the algebra for at least a single β, or if it does, that the different involutions coincide. We think we have to live with this, and take it as an important insight. This is where our approach differs the most from the existing ones where a particular fundamental symmetry commuting with the algebra is part of the structure. Still, there are two ways that we can see which permit to recover C^*-algebras out of spectral spacetimes. Note that in contrast with other approaches, these C^*-structures will be "end products": they are not fixed at the beginning.

Let us start with some notations. Given a spectral spacetime S, we will write $\mathrm{T}-\mathrm{or}(S)$ for the set of time-orientation forms. Let us define $\mathrm{Obs}(S) := \{\beta \in \mathrm{T}-\mathrm{or}(S) | \forall a \in \mathcal{A}, \pi(a)^{*_\beta} \in \pi(\mathcal{A})\}$. We call *observers* the elements of $\mathrm{Obs}(S)$. The name has two origins: the first is that in the case $S = S_{\text{can}}$, the elements of $\mathrm{Obs}(S) = \mathrm{T}-\mathrm{or}(S)$ can be seen as partitions of spacetimes into observers' trajectories (first by dualizing the form to a timelike vector field, then by integration). The second is that in the general case, each element of $\mathrm{Obs}(S)$ can be used to define its own Jordan algebra of observables. Indeed, each $\beta \in \mathrm{Obs}(S)$ defines an involution of $\pi(\mathcal{A})$, which is consequently a C^*-subalgebra of $(B(\mathcal{K}), *_\beta)$, the C^*-algebra of bounded operators on the Hilbert space $(\mathcal{K}, \langle .,. \rangle_\beta)$. The elements of $\pi(\mathcal{A})$ which are $*_\beta$-self-adjoint thus form the Jordan algebra of observables corresponding to the "observer" β. If π is injective, i.e. if the representation is faithful, we can carry the C^*-structure defined by $*_\beta$ on $\mathcal{A}$ itself. If $\mathrm{Obs}(S) \neq \emptyset$ and π is injective, in which case we say that S *is reconstructible*, there is thus at least one C^*-structure on $\mathcal{A}$. In case there are several, we can easily see that they are isomorphic. More precisely, $(\pi(\mathcal{A}), *_\beta) \simeq (\pi(\mathcal{A}), *_{\beta'})$ by $x \mapsto \beta'\beta^{-1}x(\beta'\beta^{-1})^{-1}$.

Unfortunately the set $\mathrm{Obs}(S)$ may very well be empty (examples are given in [5]). In such cases we can fall back on a kind of dual construction, which we now briefly describe[1]. We define $U(S) := \{a \in \mathcal{A} | \forall \beta \in \mathrm{T}-\mathrm{or}(S), \pi(a)^{*_\beta} = \pi(a)\}$. The elements of $U(S)$ are called *universal*

[1] More details will be given in future work. Our reference on Jordan algebras is [18].

observables. We can also define the subset $U_C(S)$ which contains the *real* universal observables, that is, those which commute with C. It is easy to see that $U(S)$ and $U_C(S)$ are both real Jordan-Banach algebras (see [18] for the definition). Clearly in the case $S = S_{\text{can}}(M, g, \sigma)$, one has $U(S) = U_C(S) = \mathcal{C}(M, \mathbb{R})$ and they are associative Jordan algebras. It can be proven that $U_C(S)$ is always associative. Every Jordan-Banach algebra gives rise to a so-called universal C^*-algebra, which in our case is just the closed associative algebra generated by $U(S)$ over the complex numbers. This algebra is commutative iff $U(S)$ is associative, in which case we can apply the first Gelfand-Naimark theorem in order to recover a topological space.

5.3. The split graph again

Let us now describe an example of spectral spacetime which generalizes to the anti-Lorentzian signature the canonical spectral triple over a finite weighted graph. We use the notations of section 3.2. We take the same algebra $\mathcal{A}$, acting by the same representation π on the same space $\mathcal{C}(\tilde{V})$, which we now denote by $\mathcal{K}$ instead of $\mathcal{H}$ in order to remind ourselves that it is now endowed with a Krein product, defined by

$$(F, G)_\omega = i \sum_{e \in E} \left(F(e, -)\overline{G(e, +)} - F(e, +)\overline{G(e, -)} \right) \tag{7}$$

The Dirac operator is:

$$D = \frac{-i}{\delta(e)} F(e, \mp) \tag{8}$$

It only differs from the euclidean Dirac operator by a factor of $-i$. The chirality is the same, but the real structure now acts off-diagonally by:

$$CF(e, \pm) = \pm \overline{F(e, \mp)} \tag{9}$$

These formula are obtained by a systematic "Wick rotation" procedure (see [5]), but we can check directly that they satisfy the axioms of an anti-Lorentzian spectral spacetime, which is moreover reconstructible. The time-orientation forms are the operators of the form

$$\beta F(e, \pm) = \pm i b(e) F(e, \mp) \tag{10}$$

where $e \mapsto b(e)$ is a positive-valued function defined on the set of edges E. This form is exact iff $b(e) = \frac{f(e^+) - f(e^-)}{\delta(e)}$ for some function $f : V \to \mathbb{R}$. The existence of this function prevents the existence of directed cycle, and it is a well-known result in graph theory that directed acyclic graphs has a so-called topological ordering ([19]). This means that the spectral spacetime just constructed is stably causal iff the transitive closure of the oriented graph (G, s, t) is a partial ordering. Note that the orientation (s, t) which played a marginal role in the euclidean case (it only served to define the representation) is now a crucial part of the structure. More precisely, a change of orientation on G induces a canonical transformation on $\mathcal{H} = \mathcal{K}$, which exchanges the endpoints of some of the edges of the split graph. This is unitary for the scalar product, but not for the Krein product (7). Hence changing the orientation of G just amounts to a change of unitary representation in the spectral triple case, but can lead to a different isomorphism class of anti-Lorentzian spectral spacetimes.

5.4. A noncommutative generalization: the split Dirac structure

We will now present a noncommutative finite-dimensional example of spectral spacetime. It is a generalization of the construction of section 5.3 obtained in the following way:

- At each vertex $v \in V$ we attach a $n = 2p$-dimensional real vector space X_v equipped with an anti-Lorentzian bilinear form g_v.
- We replace the algebra $\mathcal{A} = \mathcal{C}(V)$ by the algebra of *even* sections $a : v \mapsto a_v \in \mathbb{C}l(X_v, g_v)^0$ of the discrete Clifford bundle $\bigsqcup_{v \in V} \mathbb{C}l(X_v, g_v)$.
- We put an irreducible representation space S_v of $\mathbb{C}l(X_v, g_v)$, that is, a module of Dirac spinors, at each vertex v. The representation map is $\rho_v : \mathbb{C}l(X_v, g_v) \longrightarrow \mathrm{End}(S_v)$. We fix a spinor product $(.,.)_v$ on S_v. The future cone X_v^+ in X_v is defined to be the set of vectors $\xi \in X_v$ such that $(\psi, \rho_v(\xi)\psi)_v \geq 0$ for all $\psi \in S_v$.
- We replace $\mathcal{K} = \mathcal{C}(\tilde{V})$ by the space of sections $F : (e, \pm) \mapsto F(e, \pm) \in S_{e^\pm}$ of the discrete spinor bundle $\bigsqcup_{(e,\pm) \in \tilde{V}} S_{e^\pm}$ over the split graph $\tilde{G}$.
- The representation of the algebra on K is defined by:

$$(\pi(a)F)(e, \pm) = a(e^\pm) \cdot F(e, \pm) \tag{11}$$

  where we have written $\cdot$ to denote the action of Clifford algebra elements on spinors by way of the representation $\rho_{e^\pm}$.
- We fix parallel transport operators $h_e^+ : S_{e^-} \to S_{e^+}$, and write $h_e^- = (h_e^+)^{-1}$.
- The Krein product on K is defined to be:

$$(F, G) = \sum_{e \in E} ((F(e, +), h_e^+ G(e, -)) + (F(e, -), h_e^- G(e, +)) \tag{12}$$

- The Dirac operator D is:

$$(DF)(e, \pm) = \frac{\mp 1}{\delta(e)} \gamma_e^\pm h_e^\pm F(e, \mp) \tag{13}$$

  where $\gamma_e^\pm \in X_{e^\pm}$ is a "gamma matrix" and $\delta_e \in]0, +\infty[$.
- The chirality χ is defined by:

$$(\chi F)(e, \pm) = \chi_{e^\pm} F(e, \pm) \tag{14}$$

  where $\chi_{e^\pm}$ is the chirality operator of $\mathbb{C}l(X_{e^\pm})$.
- Finally we define $(CF)(e, \pm) = h_e^\pm C_{e^\mp} F(e, \mp)$.

The structure defined above is called *the split Dirac structure* over the weighted directed graph $G = (V, E, s, t, \delta)$. The family $(h_e^\pm)_{e \in E}$ of parallel transport operators is called the *discrete connection*. We say that the discrete connection is

- *metric* iff the h_e are Krein-unitary,
- *preserves spin* iff the diagrams

$$\begin{array}{ccc} S_{e^-} & \xrightarrow{C_{e^-}} & S_{e^-} \\ {\scriptstyle h_e^+}\downarrow & & \downarrow{\scriptstyle h_e^+} \\ S_{e^+} & \xrightarrow{C_{e^+}} & S_{e^+} \end{array}$$

  commute for all $e \in E$,

IOP Conf. Series: Journal of Physics: Conf. Series **968** (2018) 012002 doi:10.1088/1742-6596/968/1/012002

- *is Clifford* iff for all e there exists a linear map $\Lambda_e : X_{e^-} \to X_{e^+}$ such that the following diagram commutes for all $v \in X_{e^-}$:

$$\begin{array}{ccc} S_{e^-} & \xrightarrow{\;v\;} & S_{e^-} \\ {\scriptstyle h_e^+}\downarrow & & \downarrow{\scriptstyle h_e^+} \\ S_{e^+} & \xrightarrow{\;\Lambda_e(v)\;} & S_{e^+} \end{array}$$

- *preserves orientation* if $\chi_{e^\pm} h_e^\pm = h_e^\pm \chi_{e^\mp}$.

It can be shown that in the continuous case these definitions match the usual ones ([12]). In particular, if the parallel transport operators of a connection on the spinor bundle have all these properties, then they come from the spin connection. One can also prove the following theorems:

Theorem 2 *([5]) The split Dirac structure is an anti-Lorentzian spectral spacetime if and only if*

(i) The elements $\gamma_e^\pm$ satisfy

$$\gamma_e^+ = h_e^+ \gamma_e^- h_e^-$$

for all e.

(ii) The discrete connection is metric, spin and orientation preserving.

Moreover if the $\gamma_{e^\pm}$ span $X_{e^\pm}$ then the discrete connection is Clifford.

Theorem 3 *([5]) Under the conditions of theorem 2, and if in addition $n > 2$ and $(h_e)_{e\in E}$ is Clifford, then the following are equivalent:*

- *the split Dirac structure is recontructible,*
- *there exists a future-directed "covariantly constant vector field", i.e. $v \mapsto u_v \in X_v^+$ such that for all vertices v, w, u_w is the image of u_v by the product of the parallel transport operators encountered along any path leading from v to w,*
- *the discrete holonomy group fixes a timelike direction.*

This theorem allows in particular to build examples of non-reconstructible anti-Lorentzian spacetimes. These cannot be Wick rotations of spectral triples.

6. Conclusion

To end this short review, we quote a few directions for future research that we think are relevant. There are three different kinds of problems that we think are worth investigating. First, there are technical issues concerning the general theory of spectral spacetimes: finding the correct analytical conditions which would allow for a reconstruction theorem (results in [4] would probably be very helpful), generalizing the definition to the odd-dimensional case, and maybe other signatures (this last problem being probably more difficult). The second direction is to develop some aspects of the theory, namely the study of stable causality (making a bridge to Franco's approach to causality in Noncommutative Geometry, and maybe isocone theory should be possible), and the classification of the finite-dimensional case (along the lines of [20], [21]). Finally, the split Dirac structure seems interesting to study for its own sake. In particular it is possible to relate it to the discretization of the Dirac operator defined in Marcolli and van Suijlekom at least in the flat case. Is it possible to go beyond ? One could also look forward for applications in particle physics (noncommutative lattice models, also in the spirit of [22]).

IOP Conf. Series: Journal of Physics: Conf. Series **968** (2018) 012002 doi:10.1088/1742-6596/968/1/012002

References

[1] A. Strohmaier. On noncommutative and pseudo-Riemannian geometry. *J. Geom. Phys.*, 56(2):175–195, 2006.

[2] M. Paschke and A. Sitarz. Equivariant Lorentzian spectral triples. arXiv:math-ph/0611029, 2006.

[3] N. Franco. *Lorentzian approach to noncommutative geometry.* PhD thesis, University of Namur FUNDP, 2011.

[4] K. van den Dungen, M. Paschke, and A. Rennie. Pseudo-Riemannian spectral triples and the harmonic oscillator. *J. Geom. Phys.*, 73:37–55, 2013.

[5] F. Besnard. On the definition of spacetimes in noncommutative geometry: part ii. arxiv.org/abs/1611.07842, 2016.

[6] J.M. Gracia-Bondía, J.C. Várilly, and H. Figueroa. *Elements of Noncommutative Geometry.* Birkhäuser, 2001.

[7] A. Connes. Noncommutative geometry and reality. *J. Math. Phys.*, 36:6194, 1995.

[8] A. Connes. On the spectral characterization of manifolds. *J. Noncommut. Geom.*, 7:1–82, 2013.

[9] A. Connes. *Noncommutative Geometry.* Academic Press, 1994.

[10] B. Iochum, T. Krajewski, and P. Martinetti. Distances in finite spaces from noncommutative geometry. *J. Geom. Phys.*, 37:100–125, 2001.

[11] P.L. Robinson. Spinors and canonical hermitian forms. *Glasgow Mathematical Journal*, 30(3):pp 263–270, 1988.

[12] F. Besnard and N. Bizi. On the definition of spacetimes in noncommutative geometry: part i. *J. Geom. Phys.*, 123:292–309, 2018.

[13] J. Bognàr. *Indefinite inner product spaces.* Springer-Verlag, 1974.

[14] H. Baum. *Spin-Strukturen und Dirac-Operatoren über pseudoriemannschen Mannigfaltigkeiten.* PhD thesis, Sekt. Mathematik d. Humboldt-Univ., 1980.

[15] A. Bernal and M. Sànchez. Smoothness of time functions and the metric splitting of globally hyperbolic spacetimes. *Commun. Math. Phys.*, 257:43–50, 2005.

[16] M. Sánchez. Causal hierarchy of spacetimes, temporal functions and smoothness of the Geroch's splitting. a revision. *Contemporanea Matematica*, 29:127–155, 2005.

[17] C. Brouder, N. Bizi, and F. Besnard. The standard model as an extension of the noncommutative algebra of forms. arxiv.org/abs/1504.03890, 2015.

[18] E. M. Alfsen and F. W. Shultz. *Geometry of State Space of Operator Algebras.* Birkhäuser, 2003.

[19] T. Cormen, C. Leiserson, R. Rivest, and C. Stein. *Introduction to Algorithms (2nd ed.).* MIT Press and McGraw-Hill, 2001.

[20] T. Krajewski. Classification of finite spectral triples. *J. Geom. Phys.*, 28:1–30, 1998.

[21] M. Paschke and A. Sitarz. Discrete spectral triples and their symmetries. *J. Math. Phys.*, 39:6191–6205, 1998.

[22] M. Marcolli and W. D. van Suijlekom. Gauge networks in noncommutative geometry. *J. Geom. Phys.*, 75:71–91, 2014.

IOP Conf. Series: Journal of Physics: Conf. Series **968** (2018) 012003 doi:10.1088/1742-6596/968/1/012003

The annoying null boundaries

Piotr T. Chruściel and Paul Klinger

Faculty of Physics and Erwin Schrödinger Institute, University of Vienna, Boltzmanngasse 5, A1090 Wien, Austria

E-mail: piotr.chrusciel@univie.ac.at, paul.klinger@univie.ac.at

Abstract. We consider a class of globally hyperbolic space-times with "expanding singularities". Under suitable assumptions we show that no C^0-extensions across a compact boundary exist, while the boundary must be null wherever differentiable (which is almost everywhere) in the non-compact case.

1. Introduction

One of the major open questions in mathematical general relativity is the behavior of globally hyperbolic space-times when singular boundaries are approached. In particular the question of extendibility of the metric across such boundaries lies at the heart of the "cosmic censorship conjecture" [1], compare [2, 3].

In a recent important paper, Sbierski [4] has established C^0-inextendibility of the Kruskal-Szekeres extension of the Schwarzschild metric. His proof makes uses of the $SO(2)\times\mathbb{R}$ symmetry of the metric, which renders the argument unsuitable in situations where no isometries exist. (See also [5].)

It is of interest to enquire whether Sbierski's analysis can be adapted to more general spacetimes of interest, without isometries. The object of this note is to point out a class of space-times where C^0-extendibility can only happen across null boundaries, if at all. Indeed, let us consider a globally hyperbolic space-time with a differentiable Cauchy time function t covering $(0,\infty)$; we will be interested in possible extensions of M towards the past, see Definition 2.3 below. The function t determines a topological splitting $M = (0,\infty)\times\mathscr{S}$, where the slices $\{\tau\}\times\mathscr{S}$ are the level sets of t; here one travels from a slice $\{t_1\}\times\mathscr{S}$ to $\{t_2\}\times\mathscr{S}$ by following the integral curves of ∇t.

Definition 1.1. We shall say that a globally hyperbolic space-time (M,g), with a time function t as just described, contains a *globally expanding singularity towards the past* if for every open set $A\subset\mathscr{S}$ there exists a sequence t_i decreasing to zero such that the (Riemannian) diameter of $\{t_i\}\times A$ within $\{t_i\}\times\mathscr{S}$ tends to infinity as $t_i\to 0$.

As discussed in more detail in Section 6 below, the space-times constructed in [6, 7], as well as Gowdy, and various other "Asymptotically Velocity Term Dominated" (AVTD) spacetimes obtained via Fuchsian methods are of this type, and the theorems below apply.

All the boundaries that we consider in this work will be achronal, hence differentiable almost everywhere by standard arguments (cf. the beginning of Section 4).

We have:

IOP Conf. Series: Journal of Physics: Conf. Series **968** (2018) 012003 doi:10.1088/1742-6596/968/1/012003

THEOREM 1.2. *Suppose that (M, g) contains a globally expanding singularity towards the past. Then in every continuous past extension of M the boundary of ∂M is null at all its differentiability points.*

Theorem 1.2 is an immediate consequence of Theorem 3.1 below, which is proved under a condition weaker, but somewhat more involved, than that of Definition 1.1.

As such, it also holds:

PROPOSITION 1.3. *Under the hypotheses of Theorem 3.1, every spacelike hypersurface in the extension which intersects the boundary $\partial\iota(M)$ of the image $\iota(M)$ also intersects $\iota(M)$ itself.*

Some further results in the same spirit concerning general extensions are also established, see Theorems 3.3 and 3.5 below.

We suspect that null extensions cannot occur either under the hypotheses above, and thus such space-times are inextendible, but we have not been able to establish this in general. However we have:

PROPOSITION 1.4. *Under the hypotheses of Theorem 1.2, (M, g) has no extensions with a continuous metric and a compact boundary.*

The proof of Proposition 1.4 can be found in Section 5.

Recall that electrovacuum space-times with compact Cauchy horizons have been studied in [8, 9], and that, in space-times in which the metric is C^3-extendible [10, 11], for such metrics compactness of the horizon implies its differentiability. We emphasise that our arguments do not need such results.

It should be recognised that our definition of expanding space-times is tied to the choice of a time-function t, and large deformations of a good time function, if one exists, will not preserve the condition. In particular it might be very difficult to determine whether or not a given space-time, presented in a coordinate system where the conditions of Definition 1.1 are *not* met, admits a time-function which will satisfy the conditions. But our results here give some geometric meaning to the notion: space-times extendible through a compact Cauchy horizon, or past-extendible through a spacelike boundary, will not be expanding in the sense of Definition 1.1 no matter what time function is used.

2. Conventions and definitions

We use the standard definition of the Riemannian diameter of a set A in the set $B \supseteq A$, where both A and B are subsets of a Riemannian manifold (M, g)

$$\mathrm{diam}(A, B) := \sup_{x,y\in A} \inf_{\substack{\gamma:[0,1]\to B \\ \gamma(0)=x\,,\,\gamma(1)=y}} \int_0^1 |\dot\gamma(s)|_g \mathrm{d}s\,. \tag{2.1}$$

Our remaining definitions follow Sbierski [4], in particular:

DEFINITION 2.1. A C^0-*extension* of a spacetime (M, g), where M is a smooth manifold and g a Lorentzian metric, is a spacetime $(\tilde M, \tilde g)$ of the same dimension, with $\tilde M$ again a smooth manifold and $\tilde g$ a continuous Lorentzian metric, together with an isometric embedding $\iota : M \to \tilde M$ such that $\iota(M)$ is a proper subset of $\tilde M$.

The timelike futures and pasts $I^\pm$ will be defined using piecewise smooth timelike curves in both M and $\tilde M$.

We define the future/past boundary of a spacetime as in [12], namely:

IOP Conf. Series: Journal of Physics: Conf. Series **968** (2018) 012003 doi:10.1088/1742-6596/968/1/012003

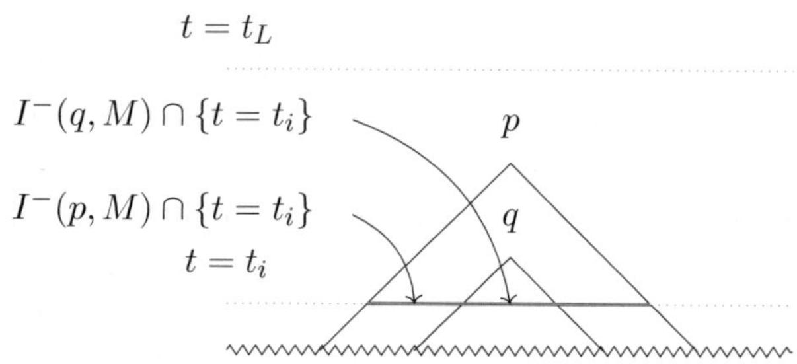

Figure 2.1: An illustration of the definition of an *expanding singularity.*

DEFINITION 2.2. The *future, respectively past, boundary of* M, denoted $\partial^+\iota(M)$, respectively $\partial^-\iota(M)$, is the set of points $p \in \partial\iota(M)$ such that there exists a future/past directed timelike curve $\gamma : [0, 1] \to \tilde{M}$ with $\gamma([0, 1)) \subset \iota(M)$ and $\gamma(1) = p$.

DEFINITION 2.3. An extension will be called *past*, respectively *future*, if $\partial^+\iota(M) = \emptyset$, respectively $\partial^-\iota(M) = \emptyset$.

We will also need a definition closely related to, but somewhat weaker than Definition 1.1:

DEFINITION 2.4. We shall say that a globally hyperbolic space-time (M, g) contains an *expanding singularity towards the past* if there exists a time function t ranging over $(0, \infty)$ and a real number $t_L > 0$ such that for all pairs of points $p \in \{t < t_L\}$ and $q \in I^-(p, M)$ there exists a sequence t_i decreasing to zero such that the diameter of $I^-(q, M) \cap \{t = t_i\}$ in $I^-(p, M) \cap \{t = t_i\}$ tends to infinity as $t_i \to 0$.

This is illustrated in Figure 2.1.

REMARK 2.5. We note that a globally expanding singularity in the sense of Definition 1.1 is expanding in the sense of Definition 2.4. For this, let us identify the spacetime M with $\mathbb{R} \times \mathscr{S}$ by flowing along the gradient ∇t of t. For a subset Ω of $\mathscr{S}$ set $\Omega(t) = \{t\} \times \Omega$, and let

$$\Omega_i(t_i) := I^-(q, M) \cap \{t = t_i\} .$$

We then have, for $j > i$,

$$\Omega_i(t_j) \subseteq I^-(q, M) \cap \{t = t_j\} \equiv \Omega_j(t_j) .$$

By its definition, given in (2.1), the diameter appearing in Definition 2.4, namely

$$\mathrm{diam}(\Omega_i(t_i), I^-(p, M) \cap \{t = t_i\}) ,$$

is bounded from below by the diameter of $\Omega_i(t_i)$ in the whole level set, $\mathrm{diam}(\Omega_i(t_i), \{t = t_i\})$. By Definition 1.1 of a globally expanding singularity the diameter of $\Omega_0(t_i)$ in $\{t = t_i\}$ diverges as t_i tends to zero, and the inclusion $\Omega_0(t_i) \subset \Omega_i(t_i)$ proves the claim.

3. Nonexistence of spacelike boundaries

We have the following result, which immediately implies Theorem 1.2:

THEOREM 3.1. *Suppose that* (M, g) *contains an expanding singularity towards the past. Then in every continuous past extension of* M *the boundary* ∂M *is null wherever differentiable.*

The following proposition gives some more information about extensions in the current context:

IOP Conf. Series: Journal of Physics: Conf. Series **968** (2018) 012003 doi:10.1088/1742-6596/968/1/012003

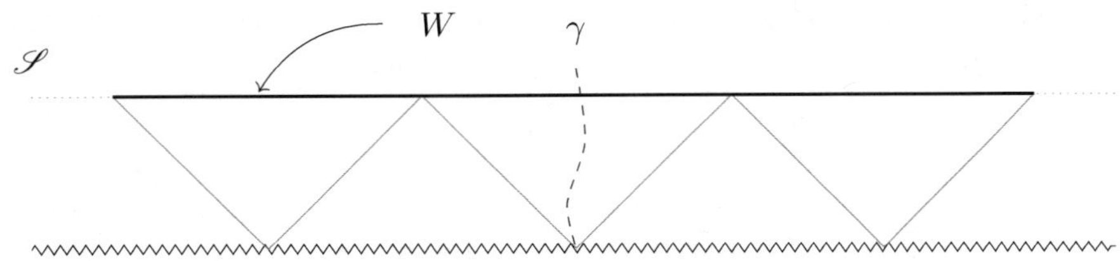

Figure 3.1: The set W appearing in the second condition of Theorem 3.5.

PROPOSITION 3.2. *Under the hypotheses of Theorem 3.1, every spacelike hypersurface in the extension which intersects $\partial^-\iota(M)$ also intersects $\iota(\{t < t_L\}) \subseteq \iota(M)$.*

For a general extension (i.e. where $\partial^+\iota(M)$ might be non-empty) we have:

THEOREM 3.3. *Suppose that (M, g) contains an expanding singularity towards the past. If the Cauchy hypersurface $\mathscr{S}$ is compact then the past boundary $\partial^-\iota(M)$ of M in every extension is null wherever differentiable.*

PROPOSITION 3.4. *Under the hypotheses of Theorem 3.3, every spacelike hypersurface in the extension which intersects $\partial^-\iota(M)$ also intersects $\iota(\{t < t_L\}) \subseteq \iota(M)$.*

Under weaker conditions on $\mathscr{S}$ we obtain a similar but more involved result:

THEOREM 3.5. *Suppose that (M, g) contains an expanding singularity towards the past. If either*

- *there exist constants $t_C > 0, C > 0$ such that for all timelike curves γ in M the intersection $I^+(\gamma, M) \cap \{t = t_C\}$ has diameter less than C,*
- *or for all timelike curves $\gamma \subset M$ the intersection*

$$W := I^+(I^-(I^+(\gamma, I^-(\mathscr{S}, M)), M), M) \cap \mathscr{S}$$

(compare Figure 3.1) is precompact.

Then:

- *every spacelike hypersurface in the extension which intersects $\partial^-\iota(M)$ also intersects $\iota(\{t < t_L\}) \subseteq \iota(M)$*
- *and for every point $p \in \partial^-\iota(M)$ there exists a neighborhood $\tilde{O} \subseteq \tilde{M}$ of p such that the hypersurface*

$$\{q \in \partial^-\iota(M) \cap \tilde{O} \,|\, I^+(q, \tilde{O}) \cap \partial^-\iota(M) = \emptyset\}, \tag{3.1}$$

which will be referred to as the futuremost part *of $\partial^-\iota(M)$ in $\tilde{O}$, is null wherever it is differentiable.*

REMARK 3.6. The first condition in Theorem 3.5 implies the second one.

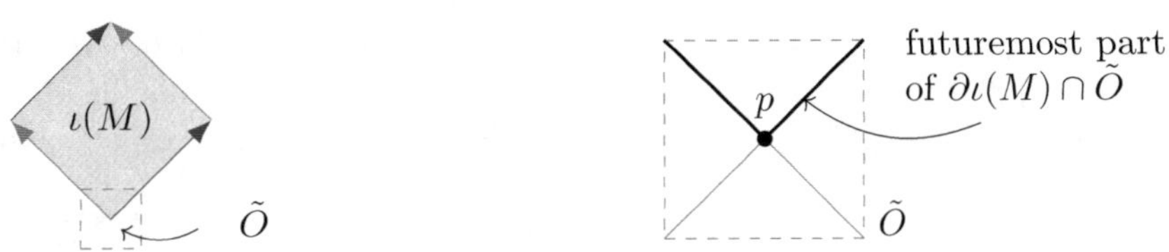

Figure 3.2: An example of a case where the futuremost part of $\partial^-\iota(M)$ in $\tilde{O}$ is not the same as $\partial^-\iota(M) \cap \tilde{O}$. The picture on the left shows the whole spacetime, with the extension given by gluing the edges according to the arrows. The picture on the right shows a neighborhood $\tilde{O}$ of the point $p \in \partial^-(M)$. Here $\partial^-\iota(M) = \partial^+\iota(M) = \partial\iota(M)$ and there is no neighborhood $\tilde{O}$ of p such that $\partial\iota(M) \cap \tilde{O}$ is achronal. The futuremost part, however, is achronal.

IOP Conf. Series: Journal of Physics: Conf. Series **968** (2018) 012003 doi:10.1088/1742-6596/968/1/012003

Figure 3.2 illustrates the difference between the *futuremost part* of $\partial^-\iota(M)$, which appears in the second consequence of Theorem 3.5, and $\partial^-\iota(M)$ itself.

4. Achronality of boundaries

In order to prove the results above, we start by showing that the relevant sets are achronal. As such, achronal sets are locally Lipschitz continuous (for every point we can find a coordinate neighborhood on which the metric is close to Minkowski, which provides a Lipschitz bound) and therefore differentiable almost everywhere by Rademacher's theorem. By achronality they are null or spacelike wherever differentiable. To finish the proofs we will need to rule out spacelike tangent planes; this will be done in the next section.

We have:

LEMMA 4.1. *The boundary of a future or past extension is an achronal topological hypersurface.*

PROOF. We assume a past extension. For a future extension, replace $I^\pm$ with $I^\mp$ in the following.

Achronality of $\partial\iota(M)$ is equivalent to $I^+(\partial\iota(M),\tilde{M}) \cap \partial\iota(M) = \emptyset$. We first show that

$$I^+(\partial\iota(M),\tilde{M}) \subseteq I^+(\iota(M),\tilde{M})\,.$$

Indeed, for every point $p \in I^+(\partial\iota(M),\tilde{M})$ there is $q \in \partial\iota(M)$ such that $p \in I^+(q,\tilde{M})$. Therefore $q \in I^-(p,\tilde{M})$ and, because I^- is open and q lies on the boundary of $\iota(M)$, there is a point $\iota(M) \ni r \in I^-(p,\tilde{M})$. This implies $p \in I^+(r,\tilde{M})$, which proves the claim.

As $\tilde{M}$ is a past extension, no future directed timelike curve leaves M, i.e. $I^+(r,\tilde{M}) \subseteq \iota(M)\ \forall r \in \iota(M)$, and therefore

$$I^+(\partial\iota(M),\tilde{M}) \subseteq \iota(M) \subseteq \tilde{M} \setminus \partial\iota(M)$$

which implies that $\partial\iota(M)$ is achronal.

To show that it is a topological hypersurface we need to show $\partial\iota(M) \cap \mathrm{edge}(\partial\iota(M)) = \emptyset$ (see [13, Section 14, Lemma 25], and note that the proof there does not use differentiability of the metric). We consider a point $p \in \partial\iota(M)$. By the above, $I^+(p,\tilde{M}) \subseteq \iota(M)$ and, as $\partial^+(\iota(M)) = \emptyset$, $I^-(p,\tilde{M}) \subset \tilde{M} \setminus \iota(M)$. Therefore every (past directed) timelike curve from $I^+(p,\tilde{M})$ to $I^-(p,\tilde{M})$ has to cross $\partial^-(M) \subseteq \partial\iota(M)$ and so $p \notin \mathrm{edge}(\partial\iota(M))$. □

REMARK 4.2. A very similar result was shown by Galloway and Ling [12, Theorem 2.6].

Next we show that every point on an achronal boundary is the limit of a timelike curve in $\iota(M)$. For a general boundary this is only true for at least one point [4, Lemma 2.17].

LEMMA 4.3. *Given a point $p \in \partial\iota(M)$ and a neighborhood $p \in \tilde{O} \subset \tilde{M}$ such that $\partial\iota(M) \cap \tilde{O}$ is achronal in $\tilde{O}$ then p is the end point of a differentiable timelike curve in M, i.e. there exists a differentiable timelike curve $\sigma : [-1,0] \to \tilde{M}$ such that $\sigma([-1,0)) \subset \iota(M)$ and $\sigma(0) = p$.*

PROOF. The curve σ might be future- or past-directed, we consider both cases here. We choose coordinates $(x^\alpha) \in (-\delta,\delta) \times (-\epsilon,\epsilon)^d$ on $\tilde{O}$ such that $p = (0,\dots,0)$, the metric is close to Minkowski and the sets $\{\pm\frac{19}{20}\delta\} \times (-\epsilon,\epsilon)^d$ are contained in $I^\pm(p,\tilde{O})$ (by choosing $\tilde{O}$ smaller if necessary).

For every $\underline{x} \in (-\epsilon,\epsilon)^d$ there is at most one $t \in (-\delta,\delta)$ such that $(t,\underline{x}) \in \partial\iota(M)$: If there were more than one then they could be connected by a (vertical) timelike curve, contradicting achronality.

As p is a boundary point there is a point $\tilde{O} \ni q \in \iota(M)$. It can be connected by a vertical curve which does not cross $\partial\iota(M)$ to one of $\{\pm\frac{19}{20}\delta\} \times (-\epsilon,\epsilon)^d$. Therefore the differentiable timelike curve $\sigma : (0,\frac{19}{20}\delta) \to \tilde{O}, \sigma(s) = (\pm s,0,\dots,0)$ lies in $\iota(M)$, as it can be connected to q by a curve which does not cross the boundary. □

IOP Conf. Series: Journal of Physics: Conf. Series **968** (2018) 012003 doi:10.1088/1742-6596/968/1/012003

REMARK 4.4. Lemma 4.1 and Lemma 4.3 together imply that if $\partial^+\iota(M) = \emptyset$ then $\partial\iota(M) = \partial^-\iota(M)$, with a similar statement obtained by reversing time-orientation.

In order to prove Theorems 3.3 and 3.5 we will need the following two Lemmas.

LEMMA 4.5. *We consider a globally hyperbolic spacetime (M, g) and a C^0-extension $(\tilde{M}, \tilde{g})$ with embedding $\iota : M \to \tilde{M}$. If there exists a spacelike Cauchy hypersurface $\mathscr{S} \subset M$ such that for all timelike curves $\gamma \subset M$ the set*

$$W := I^+(I^-(I^+(\gamma, I^-(\mathscr{S}, M)), M), M) \cap \mathscr{S}$$

is precompact in M, then for any $p \in \partial^-\iota(M)$ there exists a neighborhood $\tilde{O}$ of p such that the futuremost part of $\partial^-\iota(M)$ in $\tilde{O}$, $\{q \in \partial^-\iota(M) \cap \tilde{O} \,|\, I^+(q, \tilde{O}) \cap \partial^-\iota(M) = \emptyset\}$, is a non-empty achronal topological hypersurface.

Recall that the set W is shown in Figure 3.1.

REMARK 4.6. A simple condition that gives a precompact W is an upper bound for the diameter of $I^+(\gamma, M) \cap \mathscr{S}$ which is uniform with respect to γ.

PROOF. Let $p \in \partial^-\iota(M)$ be a point on the past boundary, γ the timelike curve approaching it, and $\mathscr{S}$ the spacelike Cauchy hypersurface as in the Lemma. As $\bar{W}$ (closure in M) is compact we can choose a neighborhood $\tilde{O}$ of p such that $\tilde{O} \cap \iota(\bar{W}) = \emptyset$. Choosing a smaller neighborhood if necessary we introduce coordinates $(x^\alpha) \in (-\delta, \delta) \times (-\epsilon, \epsilon)^d$ such that the metric is close to Minkowski, $\dot{\gamma} = \partial_{x^0}$ and $\{\pm\frac{19}{20}\delta\} \times (-\epsilon, \epsilon)^d \subseteq I^\pm(p, \tilde{O})$.

We consider the straight (in coordinates) timelike curves $\sigma_{\underline{x}} : (-\delta, \frac{19}{20}\delta] \to \tilde{O}, \; s \mapsto (s, \underline{x})$ where $\underline{x} \in (-\epsilon, \epsilon)^d$. These end in $\iota(M)$ by the construction of the coordinates. We define

$$b(\underline{x}) = \inf\{s \in (-\delta, \frac{19}{20}\delta] \mid \forall s' > s, \sigma_{\underline{x}}(s') \subset \iota(M)\}$$

and show that $b(\underline{x}) > -\delta$, i.e. that the $\sigma_{\underline{x}}$ intersect $\partial^-\iota(M)$ at least once, at $\sigma_{\underline{x}}(b(\underline{x}))$, and that

$$\sigma_{\underline{x}}(b(\underline{x})) \in \tilde{O} \setminus (I^+(p, \tilde{O}) \cup I^-(p, \tilde{O}))\,.$$

As $\mathscr{S}$ is a Cauchy hypersurface and by the definition of W, the future in $\tilde{O}$ of any point in $I_\gamma \cap \tilde{O}$, where

$$I_\gamma := \iota(I^-(I^+(\gamma, M) \cap I^-(\mathscr{S}, M), M))\,,$$

has to be contained in $\iota(M)$. In particular, as the future of a point is the future of any past directed curve ending at that point and $\gamma \cap \tilde{O} \subset I_\gamma$, $I^+(p, \tilde{O}) = I^+(\gamma, \tilde{O}) \subset \iota(M)$ and therefore

$$\sigma_{\underline{x}}(b(\underline{x})) \notin I^+(p, \tilde{O})\,.$$

We now have to show that $b(\underline{x}) > -\delta$ and $\sigma_{\underline{x}}(b(\underline{x})) \notin I^-(p, \tilde{O})$. If we assume this is false then there exists $s^- \in (b(\underline{x}), \frac{19}{20}\delta)$ such that $\sigma_{\underline{x}}(s^-) \in I^-(p, \tilde{O})$. But we also have $\sigma_{\underline{x}}(s^-) \subset I_\gamma$: The end point $\sigma_{\underline{x}}(\frac{19}{20}\delta)$ of $\sigma_{\underline{x}}$ is contained in $\{\frac{19}{20}\delta\} \times (-\epsilon, \epsilon)^d \subseteq I^+(p, \tilde{O}) = I^+(\gamma, \tilde{O})$, and therefore $\sigma_{\underline{x}}((b(\underline{x}), \frac{19}{20}\delta]) \subset I_\gamma$. Now, as $\sigma_{\underline{x}}(s^-) \subset I_\gamma$ and $p \in I^+(\sigma_{\underline{x}}(s^-), \tilde{O})$, we have $p \in \iota(M)$ by the argument in the previous paragraph, which is a contradiction.

Repeating the argument with p replaced by $\sigma_{\underline{x}}(b(\underline{x}))$ for all $\underline{x} \in (-\epsilon, \epsilon)^d$ shows that the set $\{\sigma_{\underline{x}}(b(\underline{x})) \mid \underline{x} \in (-\epsilon, \epsilon)^d\} = \{q \in \partial^-\iota(M) \mid I^+(q, \tilde{O}) \cap \partial^-\iota(M) = \emptyset\}$ is achronal in $\tilde{O}$. It is a topological hypersurface as p can't be an edge point by the properties of $b(\underline{x})$. □

Under the stronger assumption that the Cauchy hypersurface $\mathscr{S}$ is compact, we obtain the following simpler result.

IOP Conf. Series: Journal of Physics: Conf. Series **968** (2018) 012003 doi:10.1088/1742-6596/968/1/012003

LEMMA 4.7. *We consider a globally hyperbolic spacetime (M, g) and a C^0-extension $(\tilde{M}, \tilde{g})$ with embedding $\iota : M \to \tilde{M}$. If there exists a compact spacelike Cauchy hypersurface of M then $\partial^- \iota(M)$ is a locally achronal topological hypersurface.*

PROOF. Let $\mathscr{S}$ be the compact spacelike Cauchy hypersurface. As any subset of a compact set is precompact by definition, Lemma 4.5 applies. What remains to be shown is that the set $\{q \in \partial^- \iota(M) \cap \tilde{O} \,|\, I^+(q, \tilde{O}) \cap \partial^- \iota(M) = \emptyset\}$ is actually the full $\partial^- \iota(M) \cap \tilde{O}$, i.e. that there are no additional points in $\partial^- \iota(M)$ below it.

We choose a neighborhood $\tilde{O}$ and coordinates as in the proof of Lemma 4.5, but with $\tilde{O} \cap \mathscr{S} = \emptyset$. Now the future in $\tilde{O}$ of any point $q \in \partial^- \iota(M) \cap \tilde{O}$ lies in $\iota(M)$.

If there was a point $q \in \partial^- \iota(M) \cap \tilde{O}$ such that $\partial^- \iota(M) \ni p \in I^+(q, \tilde{O}) \cap \partial^- \iota(M)$ then p would lie in $\iota(M)$. □

5. Proofs of the main theorems

The proofs depend on the following Lemma, which is a slight variation of a result of Sbierski [14]. We use his notation for the sets

$$C_a^- := \{0 \neq X \in \mathbb{R}^{d+1} | \frac{< X, e_0 >_{\mathbb{R}^{d+1}}}{|X|_{\mathbb{R}^{d+1}}} < -a\},$$

where $0 < a < 1$, $< .,. >_{\mathbb{R}^{d+1}}$ is the Euclidean scalar product in $\mathbb{R}^{d+1}$ and $|.|_{\mathbb{R}^{d+1}}$ is the Euclidean norm. The C_a^- are cones of vectors with angle less than $\cos^{-1}(a)$ to the x^0 axis with the tip of the cone pointing up.

LEMMA 5.1. *We consider a spacetime (M, g) with extension $(\tilde{M}, \tilde{g})$. Given a neighborhood $\tilde{O}$ of a point on the boundary of $\iota(M)$, a point $p \in \tilde{O} \cap \iota(M)$ and a chart $\tilde{\psi} : \tilde{O} \to (-\delta, \delta) \times (-\epsilon, \epsilon)^d$ such that*

(i) ∂_{x_0} is timelike,

(ii) $|\tilde{g}_{\alpha\beta} - \eta_{\alpha\beta}| < \nu$ where η is the Minkowski metric and $1/2 > \nu > 0$ a constant such that $\forall a \in (-\delta, \delta) \times (-\epsilon, \epsilon)^d$, $\tilde{\psi}^{-1}(a + C_{5/6}^-) \subseteq I^-(\tilde{\psi}^{-1}(a), \tilde{O}) \subseteq \tilde{\psi}^{-1}(a + C_{5/8}^-)$,

(iii) $\tilde{\psi}^{-1}(\{\tilde{x}^0 < -\frac{1}{10}\delta\}) \subseteq \tilde{M} \setminus I^-(\tilde{\psi}^{-1}(\{\frac{19}{20}\delta\} \times (-\epsilon, \epsilon)^d), \iota(M))$,

(iv) $I^-(p, \iota(M)) \subset \tilde{O}$,

(v) $\tilde{\psi}^{-1}(\{\frac{19}{20}\delta\} \times (-\epsilon, \epsilon)^d) \subseteq \iota(I^+(p, \tilde{O}))$,

there exist $q \in I^-(p, \iota(M)) \cap \tilde{O}$ and a constant $0 < C_d < \infty$ such that for all Cauchy hypersurfaces $\mathscr{S}$ of M the distance in $I^-(p, \iota(M)) \cap \iota(\mathscr{S})$ of any two points in $I^-(q, \iota(M)) \cap \iota(\mathscr{S})$ is bounded above by C_d.

PROOF. We choose $q \in I^-(p, \tilde{O}) \cap \iota(M)$ such that

$$(\tilde{\psi}(q) + C_{5/8}^-) \cap \{\tilde{x}^0 > -\delta/10\} \subset (\tilde{\psi}(p) + C_{5/6}^-) \cap \{\tilde{x}^0 > -\delta/10\},$$

i.e. such that the past of q in M lies completely inside a (Euclidean) cone contained in the past of p.

We assume $q \in \iota(I^+(\mathscr{S}, M))$, as otherwise $I^-(\iota^{-1}(q), M) \cap \mathscr{S} = \emptyset$ and there is nothing to show. By property (iii) of the chart $\tilde{\psi}$ there exists a function $(-\epsilon, \epsilon)^d \to (-\delta, \delta)$, $\underline{x} \mapsto L_{\underline{x}}$ such that the timelike curves $\sigma_{\underline{x}} : (L_{\underline{x}}, \delta) \to M, \sigma_{\underline{x}}(s) = \iota^{-1}(\tilde{\psi}^{-1}(s, \underline{x}))$ are past inextendible in M. As q lies in the past of p, property (v) holds for q as well. This implies that the curves $\sigma_{\underline{x}}$ intersect $I^+(\mathscr{S}, M)$ and therefore intersect $\mathscr{S}$ exactly once, say at $s = f(\underline{x})$. We take this as the definition of $f : (-\epsilon, \epsilon)^d \to (-\delta, \delta)$.

We first show that f is smooth: As $\tilde{\psi}(\iota(\mathscr{S}) \cap \tilde{O})$ is a smooth submanifold, there exists for every point $(f(\underline{x}_0), \underline{x}_0)$ a neighborhood W and a smooth submersion $g : W \to \mathbb{R}$ such that $\tilde{\psi}(\iota(\mathscr{S}) \cap \tilde{O}) \cap W = \{g = 0\}$. As $\mathscr{S}$ is a Cauchy hypersurface no timelike vector can be tangent to it. Therefore $\partial_0 g|_{(f(\underline{x}_0), \underline{x}_0)} \neq 0$ and by the implicit function theorem there exists a smooth function $h : (-\epsilon, \epsilon)^d \supseteq V \to (-\delta, \delta)$, where V is a neighborhood of $\underline{x}_0$, such that $g(h(\underline{x}), \underline{x}) = 0$. Thus $f|_V = h$ and therefore f is smooth.

The next step is to show that $|\partial_i f|$ is bounded by a positive constant in $\tilde{O}$ for all i. As vectors tangent to $\mathscr{S}$ cannot be timelike we obtain the inequality

$$0 \leq \tilde{g}((\partial_i f)\partial_0 + \partial_i, (\partial_i f)\partial_0 + \partial_i) = (\partial_i f)^2 \tilde{g}_{00} + 2(\partial_i f)\tilde{g}_{0i} + \tilde{g}_{ii} \,. \tag{5.1}$$

By property (ii), $\tilde{g}_{00} < -1/2$, $\tilde{g}_{ii} > 1/2$, and $\tilde{g}_{0i} < 1/2$ and therefore this inequality is only satisfied for $(\partial_i f)_- \leq (\partial_i f) \leq (\partial_i f)_+$ where $(\partial_i f)_\pm$ are the values where equality holds in (5.1), i.e.

$$(\partial_i f)_\pm = \frac{-\tilde{g}_{0i} \mp \sqrt{(\tilde{g}_{0i})^2 - \tilde{g}_{ii}\tilde{g}_{00}}}{\tilde{g}_{00}} \,.$$

Again using property (ii), we see that $(\partial_i f)_\pm$, and therefore also $(\partial_i f)$, are bounded by a constant independent of f.

We define $\omega : (-\epsilon, \epsilon)^d \to (-\delta, \delta) \times (-\epsilon, \epsilon)^d$, $\omega(\underline{x}) = (f(\underline{x}), \underline{x})$. This parameterizes a smooth submanifold $\tilde{S}$ which is isometric to an open subset of $\mathscr{S}$ in M by $\iota^{-1} \circ \tilde{\psi}^{-1}$. We denote by $\bar{g}$ the metric induced on $\tilde{S}$ by $\tilde{g}$. The components of $\bar{g}$ are

$$\bar{g}_{ij} = \tilde{g}_{\mu\nu} \frac{\partial \omega^\mu}{\partial x_i} \frac{\partial \omega^\nu}{\partial x_j} = \tilde{g}_{00} \frac{\partial f}{\partial x_i} \frac{\partial f}{\partial x_j} + \tilde{g}_{0j} \frac{\partial f}{\partial x_i} + \tilde{g}_{i0} \frac{\partial f}{\partial x_j} + \tilde{g}_{ij} \,.$$

As $|\tilde{g}_{\mu\nu}|$ and $\partial_i f$ are bounded by the above, we have $|\bar{g}_{ij}| < C_{\bar{g}}$ for a positive constant $C_{\bar{g}}$.

We now consider two points $r, s \in I^-(q, \iota(M)) \cap \iota(M)$ as in the lemma. As $I^-(q, M) \subset \tilde{O}$ there exist $\underline{x}, \underline{y} \in (-\epsilon, \epsilon)^d$ such that $\omega(\underline{x}) = \tilde{\psi}(\iota(r))$ and $\omega(\underline{y}) = \tilde{\psi}(\iota(s))$. The length of the straight line $\sigma : [0, 1] \to (-\epsilon, \epsilon)^d$, $\sigma(\ell) = \underline{x} + \ell(\underline{y} - \underline{x})$ is given by

$$\begin{aligned}
L(\sigma) &= \int_0^1 \sqrt{\bar{g}(\dot{\sigma}(\ell), \dot{\sigma}(\ell))} \, \mathrm{d}\ell \\
&= \int_0^1 \sqrt{\sum_{i,j=1}^d (\underline{y}_i - \underline{x}_i)\bar{g}_{ij}(\sigma(\ell))(\underline{y}_j - \underline{x}_j)} \, \mathrm{d}\ell \\
&\leq \int_0^1 \sqrt{\sum_{i,j=1}^d 2\epsilon \cdot C_{\bar{g}} \cdot 2\epsilon} \, \mathrm{d}\ell \\
&= 2\epsilon d \sqrt{C_{\bar{g}}} \,.
\end{aligned}$$

As $I^-(q, \iota(M)) \cap \tilde{O} \subset \tilde{\psi}^{-1}(\tilde{\psi}(p) + C^-_{5/6})$ the curve σ is contained in $I^-(p, M) \cap \mathscr{S}$. As the distance in $I^-(p, M) \cap \mathscr{S}$ between r and s is defined as the infimum over the length of all piecewise smooth curves connecting them, and $\iota^{-1} \circ \tilde{\psi}^{-1} \circ \sigma$ is one such curve, this completes the proof. □

We are now ready to prove Theorem 3.1.

Proof of Theorem 3.1. By Lemma 4.1 the boundary $\partial\iota(M)$ of the extension is achronal, hence differentiable almost everywhere, and by Lemma 4.3 we have for every $p_\partial \in \partial\iota(M)$ a timelike curve $\gamma : [0, 1] \to \tilde{M}$ such that $\gamma([0, 1)) \subset \iota(M)$ and $\gamma(1) = p_\partial$.

IOP Conf. Series: Journal of Physics: Conf. Series **968** (2018) 012003 doi:10.1088/1742-6596/968/1/012003

We assume that $\partial\iota(M)$ is spacelike at a point p_∂ at which it is differentiable and establish a contradiction. We choose a neighborhood $\tilde{O}$ of p_∂ such that $t < t_L$ in $\tilde{O} \cap \iota(M)$, where t_L is the constant appearing in the Definition 2.4 of expanding singularity. This is possible as $\partial\iota(M)$ is achronal: In any neighborhood of p_∂, either $t < t_L$ or $\{t = t_L\}$ is some achronal set lying above $\partial\iota(M)$ and we can find a smaller neighborhood of p_∂ which doesn't intersect it.

We choose coordinates $\tilde{x}^i$ on $\tilde{O}$ such that $\tilde{x}^i(p_\partial) = 0$, $\tilde{g}(p_\partial) = \eta$ (where η is the Minkowski metric) and $|\tilde{g}(x)_{\alpha\beta} - \eta_{\alpha\beta}| < \nu$ for all $x \in \tilde{O}$ and ν such that condition 2 of Lemma 5.1 is satisfied.

We perform a Lorentz boost to transform the normal vector of $\partial\iota(M)$ at p_∂ to ∂_{x^0} and, by choosing a smaller neighborhood $\tilde{O}$ if necessary, ensure that $\partial\iota(M)$ is almost horizontal, i.e. $\partial\iota(M) \cap \tilde{O} \subset \{-\delta/10 < \tilde{x}^0 < +\delta/10\}$ and therefore condition 3 is satisfied.

We choose a point $p \in \tilde{O} \cap \iota(M)$ such that $I^-(p, \iota(M)) \subset \tilde{O}$, satisfying condition 4. Finally, by choosing a smaller neighborhood, we can satisfy the remaining condition 5.

Applying Lemma 5.1 we obtain, for a point $q \in I^-(p, \iota(M))$ and for all Cauchy hypersurfaces $\mathscr{S}$, an upper bound for the distance in $I^-(p, \iota(M)) \cap \mathscr{S}$ of any two points $r, s \in I^-(q, M) \cap \mathscr{S}$.

This is a contradiction to Definition 2.4 of expanding singularities. □

Proof of Proposition 3.2. To prove that there exists no spacelike hypersurface $\Sigma_{p_\partial} \subset \tilde{M} \setminus \iota(M)$ such that $p_\partial \in \Sigma_{p_\partial}$ we use the same argument, but choose coordinates in $\tilde{O}$ such that Σ_{p_∂} instead of $\partial\iota(M)$ is almost horizontal. As Σ_{p_∂} is spacelike, the part of $I^-(p, \tilde{M})$ lying above Σ_{p_∂} is entirely contained in $\tilde{O}$, we denote it by $I_{\Sigma_{p_\partial},p}$. As any past directed curve from p has to cross Σ_{p_∂} before leaving $\tilde{O}$ we have $I^-(p, \iota(M)) \subseteq I_{\Sigma_{p_\partial},p}$ and therefore we obtain a contradiction as before. □

The proofs of Theorems 3.3 and 3.5 proceed in a very similar way:

Proof of Theorems 3.3 and 3.5. By Lemma 4.7 the conditions of Theorem 3.3 imply that $\partial^-\iota(M)$ is a locally achronal topological hypersurface. Similarly, under the conditions of Theorem 3.5, Lemma 4.5 implies that the "locally futuremost part" of the past boundary $\partial^-\iota(M)$ is achronal, i.e. for every $p \in \partial^-\iota(M)$ there exists a neighborhood U of p such that

$$\{z \in \partial^-\iota(M) \cap U \mid I^+(z, U) \cap \partial^-\iota(M) = \emptyset\}$$

is an achronal topological hypersurface in U.

By the definition of $\partial^-\iota(M)$ we have for every $p \in \partial^-\iota(M)$ a timelike curve γ ending at p. The proof now proceeds analogously to that of Theorem 3.1 with $\partial\iota(M)$ replaced by $\partial^-\iota(M)$ in the case of Theorem 3.3 and with the "locally futuremost part" of $\partial^-\iota(M)$ in the case of Theorem 3.5.

The argument excluding the existence of a spacelike hypersurface $\Sigma_{p_\partial} \subset \tilde{M} \setminus \iota(M)$ such that $p_\partial \in \Sigma_{p_\partial}$ follows the proof of Proposition 3.2, except that to guarantee that $I^-(p, \iota(M))$ lies above Σ_{p_∂} we only need $\Sigma_{p_\partial} \cap \iota(\{t < t_L\}) = \emptyset$ as $t < t_L$ in $I^+(\partial^-\iota(M), \tilde{O})$. □

We pass now to the

Proof of Proposition 1.4. In contrast to the proofs above, if the boundary is null we cannot use the pasts of points in M to identify subsets of Cauchy hypersurfaces in neighborhoods in $\tilde{M}$ with those in M. Instead we consider the total diameter of the (compact) Cauchy hypersurface. We first need to ensure that neighborhoods of the boundary contain the whole (embedded) Cauchy hypersurface $\mathscr{S}_s := \{t = s\}$, for s sufficiently small.

By the proof of Theorem 3.1 in [15] there exists a constant $\delta > 0$ such that for all $0 < s \leq \delta$ there is a map $\psi_s : \tilde{M} \to \tilde{M}$, the flow of a continuous timelike vector field on $\tilde{M}$, such that

IOP Conf. Series: Journal of Physics: Conf. Series **968** (2018) 012003 doi:10.1088/1742-6596/968/1/012003

$\psi_s(\partial\iota(M)) = \iota(\mathscr{S}_s)$ and these are compact. In [15] it is assumed that the extended manifold is at least C^3, but this can be relaxed in our case: The results of [15, Lemma 3.2] hold by our Lemma 4.1. Note that [15, Lemma 2.1] and [15, Lemma 3.3] require only differentiabilty of the manifold, not the metric. The rest of the proof takes place in (M, g) which is smooth in any case. In addition, [15, Theorem 3.1] assumes that the boundary itself is C^1 but this can similarly be relaxed as long as the Cauchy hypersurfaces $\mathscr{S}_t$ are C^1.

By Lemma 4.1 and Lemma 4.3 the boundary $\partial\iota(M) = \partial^-\iota(M)$ of the extension is achronal. We choose for each point $p \in \partial\iota(M)$ an open neighborhood in $\tilde{M}$ and coordinates such that the metric is close to Minkowski. As $\partial\iota(M)$ is compact we can find a finite subcover $\{O_i\}$. For each i there exists a constant δ_i such that for each $p \in \bar{O}_i \cap \partial\iota(M)$ and all $0 < s \leq \delta_i$, $\psi_s(p) \in \bar{O}_i$. As there are only finitely many i we can set $\delta_{min} = \min\{\delta_i\}$ and obtain $\mathscr{S}_s = \psi_s(\partial\iota(M)) \subset \bigcup_i \bar{O}_i$ for all $0 < s \leq \delta_{min}$.

The distance in $\mathscr{S}$ between any two points of a Cauchy hypersurface $\mathscr{S}$ is bounded above in each O_i by a constant independent of $\mathscr{S}$ by a similar argument as in the proof of Lemma 5.1. As $I^+(\partial\iota(M), \tilde{M}) \subset \iota(M)$ and the Cauchy hypersurfaces $\mathscr{S}_t$ are achronal in M they can intersect O_i only once (i.e. $\iota(\mathscr{S}_t) \cap O_i$ is connected), and therefore the total diameter of $\mathscr{S}_t$ is bounded. This contradicts the Definition 1.1 of a globally expanding singularity. □

6. Examples

We will use the following Lemma to show that the examples below contain an expanding singularity towards the past, as required by Theorem 3.1:

LEMMA 6.1. *We consider a globally hyperbolic spacetime (M, g) of dimension $n+1$ with a Cauchy time function $t : M \to (0, \infty)$ such that $M = (0, \infty) \times \mathscr{S}$. Suppose that there exists $t_L > 0$ such that the subset $\{t < t_L\} \subset M$ can be covered by charts of the form $(0, t_L) \times U$, for some open subset $U \subseteq \mathscr{S}$, in which the metric takes the form*

$$g = g_{00}(t, x)\mathrm{d}t^2 + g_{ij}(t, x)\mathrm{d}x^i\mathrm{d}x^j \,, \quad g_{00} < 0 \,,$$

and satisfies

$$g_{11}(t, x) \xrightarrow{t \to 0} \infty \,, \quad g_{ij}(t, x) \xrightarrow{t \to 0} 0 \quad \text{for} \quad (i, j) \neq (1, 1) \,, \tag{6.1}$$

$$\text{and either} \quad g_{1i} = 0 \quad \text{for} \quad i \neq 1 \quad \text{or} \quad \frac{g_{mi}g_{mj}}{g_{mm}g_{ij}} \xrightarrow{t \to 0} 0 \quad \text{for} \quad i, j < m \tag{6.2}$$

uniformly on compact subsets in x. Assume moreover that either

- *$\mathscr{S}$ is compact and $\{t < t_L\}$ is covered by a single chart as above,*
- *or for every $p \in \{t < t_L\} \subset M$ there exists a chart as above which contains $I^-(p, M)$ and a compact set $K_p \subset \mathscr{S}$ such that $I^-(p, M) \subset (0, t_\sigma) \times K_p$.*

Then (M, g) contains an expanding singularity towards the past.

PROOF. We need to show that for every $p \in \{t < t_L\}$ and $q \in I^-(p, M)$ there exist a sequence t_i decreasing to zero such that the diameter of $I^-(q, M) \cap \{t = t_i\}$ in $I^-(p, M) \cap \{t = t_i\}$ tends to infinity as $t_i \to 0$.

If the second condition in (6.2) holds, we start by defining an orthonormal frame $\{\theta^m\}$, $g_{ij}dx^i dx^j = \delta_{ij}\theta^i\theta^j$ on each slice by

$$\theta^m = \sqrt{h^m_{mm}}(dx^m + \sum_{i=1}^{m-1} \frac{h^m_{mi}}{h^m_{mm}} dx^i) \,,$$

IOP Conf. Series: Journal of Physics: Conf. Series **968** (2018) 012003 doi:10.1088/1742-6596/968/1/012003

where

$$h^{m-1} = h^m - (\theta^m)^2 \quad \text{and} \quad h^n_{ij} := g_{ij} \,.$$

The tensor field h^{m-1} is positive definite on the subspace spanned by $\{\partial_1, \ldots, \partial_{m-1}\}$: Indeed, if there were a vector $X = X^1\partial_1 + \ldots X^{m-1}\partial_{m-1}$ such that $h^{m-1}(X,X) \leq 0$ then we could choose X^m so that $\theta^m(X + X^m\partial_m) = 0$. Setting $0 \neq Y := X + X^m\partial_m$, this gives $h^{m-1}(Y,Y) = h^{m-1}(X,X) \leq 0$ and $h^m(Y,Y) = h^{m-1}(X,X) \leq 0$, giving a contradiction to the positive definiteness of g.

Writing h^{m-1} in terms of h^m we obtain

$$h^{m-1} = \sum_{i,j=1}^{m-1} \left(h^m_{ij} - \frac{h^m_{mi}h^m_{mj}}{h^m_{mm}} \right) dx^i dx^j \,.$$

One checks that $h^{m-1}_{ij}/h^m_{ij} \to 1$ follows from (6.2), which further implies $h^m_{ij}/g_{ij} \to 1$ for $i,j \leq m$. By construction we have $\theta^1 = \alpha(t,x)dx^1$, with $\alpha(t,x)/\sqrt{g_{11}(t,x)} \to 1$, and with all convergences uniform on compact subsets.

We now choose $k_0 \in \mathbb{N}$ and $\epsilon > 0$ such that the sequences of points

$$y^k_1 = (1/k, x^1(q) - \epsilon, x^2(q), \ldots) \text{ and } y^k_2 = (1/k, x^1(q) + \epsilon, x^2(q), \ldots)$$

fulfill $y^k_1, y^k_2 \in I^-(q,M)$ for all $k \in \{k_0, k_0+1, \ldots\}$. The distance between y^k_1 and y^k_2, and therefore the diameter of $I^-(q,M) \cap \{t = 1/k\}$, in $I^-(p,M) \cap \{t = 1/k\}$ is bounded from below by

$$d_{\{1/k\}\times K_p}(y^k_1, y^k_2) = \inf_\sigma \int_{-\epsilon}^{\epsilon} \sqrt{\sum_{i=1}^n (\theta^i(\dot\sigma))^2 \mathrm{d}s} \geq \inf_\sigma \int_{-\epsilon}^{\epsilon} \sqrt{(\theta^1(\dot\sigma))^2}\mathrm{d}s = \inf_\sigma \int_{-\epsilon}^{\epsilon} |\alpha(t,x)\dot\sigma^1|\mathrm{d}s \,,$$

where the infimum is taken over curves $\sigma : [-\epsilon, \epsilon] \to \{1/k\} \times K_p$ (with $K_p := \mathscr{S}$ in the case where $\mathscr{S}$ is compact), such that $\sigma(-\epsilon) = y^k_1$ and $\sigma(\epsilon) = y^k_2$. If the first condition holds in (6.2) we obtain the same expression with $\alpha := g_{11}$.

We have

$$\begin{aligned} \lim_{k\to\infty} d_{\{1/k\}\times K_p}(y^k_1, y^k_2) &= \lim_{k\to\infty} \inf_\sigma \int_{-\epsilon}^{+\epsilon} |\alpha(\sigma(s))\dot\sigma^1(s)|\mathrm{d}s \\ &> \lim_{k\to\infty} \left(\min_{x\in K_p} \alpha(1/k, x) \right) \inf_\sigma \int_{-\epsilon}^{+\epsilon} |\dot\sigma^1(s)|\mathrm{d}s \,, \end{aligned}$$

where the minimum diverges by the uniform divergence of α on K_p and the integral is independent of k and positive as $\int_{-\epsilon}^{+\epsilon} \dot\sigma^1(s)\mathrm{d}s = 2\epsilon$. □

REMARK 6.2. Lemma 6.1 directly generalizes to the case where the expanding direction rotates within the $t = \text{const}$ hypersurfaces. Indeed, let $\{t_k\}$ be a sequence of times and $\{X_k\}$ a sequence of associated expanding directions (constant in x unit vectors w.r.t. the Euclidean metric δ_{ij}) in each hypersurface $\mathscr{S}_k := \{t = t_k\}$. A spatial rotation for each t_k gives $X_k = \partial_{x^1}$ and the conditions on the metric are as before in these adapted coordinates. This is relevant for the case of Bianchi IX and BKL singularities discussed in Sections 6.4 and 6.5 below.

6.1. AVTD metrics without symmetries

These solutions, constructed in [6, 7], take the form

$$\mathrm{d}s^2 = -e^{-2\sum_{a=1}^{3}\beta^a}\mathrm{d}\tau^2 + \sum_{a=1}^{3} e^{-2\beta^a}\mathcal{N}^a{}_i\mathcal{N}^a{}_j\,\mathrm{d}x^i\mathrm{d}x^j\,, \tag{6.3}$$

with β^a and $\mathcal{N}^a{}_i$, $i, a \in \{1, 2, 3\}$, depending on all coordinates τ, x^i and behaving asymptotically as

$$\beta^a = \beta^a_\circ + \tau p^a_\circ + O(e^{-\tau\nu}) \quad \text{and} \quad \mathcal{N}^a{}_i = \delta^a_i + O(e^{-2\tau(p^i_\circ - p^a_\circ)})\,, \tag{6.4}$$

with $\nu > 0$. They are parameterized by freely prescribable analytic functions $\beta^2_\circ, \beta^3_\circ$ and $P_\circ{}^2{}_1$ depending on all space coordinates and two analytic functions $p^2_\circ$ and $p^3_\circ$ depending on all space coordinates which are free except for the inequalities

$$0 < p^2_\circ < (\sqrt{2} - 1)p^3_\circ\,. \tag{6.5}$$

The function $P_\circ{}^2{}_1$ does not appear in the asymptotic expansion (6.4) but influences lower order terms in the expansion of $\mathcal{N}^a{}_i$.

The remaining exponent $p^1_\circ$ is given by

$$p^1_\circ = -\frac{p^2_\circ p^3_\circ}{p^2_\circ + p^3_\circ} < 0\,, \tag{6.6}$$

i.e. x^1 is the expanding direction.

The solutions approach a curvature singularity as $\tau \to \infty$.

The construction of these spacetimes in [7] was done in a purely local manner, regardless of the topology of the solutions. Here we will assume that the spatial topology is compact.

We replace the time coordinate τ with $t = -\log\tau$, giving, with the asymptotic expansion (6.4) inserted,

$$\mathrm{d}s^2 = -t^{2\sigma_{p_\circ}-2}e^{-2\sigma_{\beta_\circ}}(1+O^\nu)\mathrm{d}t^2 + \sum_{a=1}^{3} t^{2p^a_\circ}e^{-2\beta^a_\circ}(1+O^\nu)(\delta^a_i + O(t^{2(p^i_\circ - p^a_\circ)}))(\delta^a_j + O(t^{2(p^j_\circ - p^a_\circ)}))\mathrm{d}x^i\mathrm{d}x^j \tag{6.7}$$

where $O^\nu = O(t^\nu)$, $\sigma_{p_\circ} = p^1_\circ + p^2_\circ + p^3_\circ$, and $\sigma_{\beta_\circ} = \beta^1_\circ + \beta^2_\circ + \beta^3_\circ$. We see directly that $g_{11} \to \infty$ and $g_{ij} \to 0$, satisfying (6.1). From (6.7) we find that the metric component g_{ij} is of the same order as the faster decaying one of g_{ii} and g_{jj} and that $g_{ii}/g_{jj} \to 0$ for $i > j$. This implies that (6.2) is also satisfied, i.e. the solutions are of the form required in Lemma 6.1.

6.2. T^3 Gowdy

The T^3 Gowdy spacetimes have metrics of the form

$$g = e^{(\tau-\lambda)/2}(-e^{-2\tau}\mathrm{d}\tau^2 + \mathrm{d}\theta^2) + e^{-\tau}(e^P\mathrm{d}\sigma^2 + 2e^PQ\mathrm{d}\sigma\mathrm{d}\delta + (e^PQ^2 + e^{-P})\mathrm{d}\delta^2)\,,$$

where λ, P and Q are functions of τ and θ and the singularity is approached as $\tau \to \infty$. The $\tau = \text{const}$ slices are toroidal and therefore compact. Ringström showed [16] that generic Gowdy spacetimes asymptotically behave as follows:

$$P(\tau,\theta) = v_a(\theta)\tau + \phi(\theta) + o(1), \qquad Q(t,\theta) = q(\theta) + o(1), \qquad \lambda = v_a(\theta)^2\tau + o(\tau)$$

where $0 < v_a(\theta) < 1$ and the lower order terms converge uniformly. Therefore the $g_{\theta\theta}$ component of the metric diverges towards the singularity while all other space components converge to zero. By redefining the time coordinate as $t := -\ln\tau$ the metric can be brought to the form required in Lemma 6.1: since the off-diagonal components $g_{\theta i}$ vanish, so (6.2) is satisfied.

IOP Conf. Series: Journal of Physics: Conf. Series **968** (2018) 012003 doi:10.1088/1742-6596/968/1/012003

6.3. Further AVTD spacetimes

Using Fuchsian methods, asymptotically Kasner-like spacetimes without symmetries have been constructed in the presence of various matter fields and in vacuum (either for spacetime dimension higher than 10, or the ones described in section 6.1 which exist for a restricted set of asymptotic data). The constructions generally start by defining a reduced evolution system, the "velocity term dominated" (VTD) or "Kasner-like" system, which does not include spatial derivatives, and then using Fuchs-type theorems to show that solutions of the full Einstein equations which approach these exist. These theorems guarantee a convergence which is uniform on compact subsets, as required by Lemma 6.1. Assuming that the spatial manifold is compact, the only things left to verify are the conditions (6.1), (6.2)

In the case of stiff fluid or scalar field matter there is no expanding direction and Lemma 6.1 does not apply [17, 18].

In the case of ≥ 10 dimensional vacuum, there is at least one expanding direction [18]. These solutions can be constructed using the same approach as those in section 6.1, leading to the same behavior of the $\mathcal{N}^a_{\ i}$ and therefore also satisfy (6.2). Lemma 6.1 is applicable if the spatial manifold is compact.

There are various results on general (non-Gowdy) T^2 symmetric spacetimes, the most general of which assumes the so-called "half-polarization" condition [19, 20]. These have one expanding direction, satisfy (6.2), and are spatially compact.

6.4. "Mixmaster" Bianchi IX spacetimes

The Bianchi models are homogeneous, but generically anisotropic, spacetimes, which are divided into types according to the structure constants of their Killing vector fields.

As shown by Ringström [21], generic solutions of type IX have at least three α-limit points on the "Kasner circle", i.e. they approach at least three different Kasner metrics arbitrarily closely as the singularity is approached. It is conjectured that generic α-limit sets contain an infinite number of points on the Kasner circle, and that the dynamics approaches that of the discrete "Kasner map", which shows chaotic behavior [22]. One would naively expect to be able to choose a sequence of times and directions as in Remark 6.2. However, it has been pointed out to us by Hans Ringström (private communication) that this expectation is incorrect, and that with some work one can infer from [21] that the space-diameter of the surfaces of homogeneity in all Bianchi IX vacuum models approaches zero as the singularity is approached.

6.5. Generic spacelike singularities in the context of the BKL conjecture

The BKL conjecture states, roughly speaking, that generic spacelike singularities behave at each spatial point as a "Mixmaster" Bianchi IX solution.

There are various heuristic arguments supporting this conjecture. Using the so-called "cosmological billiards formalism", as described in [23], the metric takes the form (6.3), with the same behavior of the $\mathcal{N}^a_{\ i}$ as in (6.4), and with the $\beta^a(\tau, x)$ now *not* showing linear behavior in τ, but rather a sequence of approximately linear phases (so-called Kasner epochs) connected by "bounces" off increasingly sharp potential walls. The β^a are expected to be unbounded towards the singularity ($\tau \to \infty$), but they might well be bounded from below, and therefore it is not clear (and perhaps unlikely, given the Bianchi IX result mentioned above) whether one can choose a sequence of times and directions as described in Remark 6.2.

It is expected that such solutions will develop particle horizons (this is sometimes referred to as "asymptotic silence") [24]. This would imply that the second option in Lemma 6.1 is fulfilled, i.e. $I^-(p, M) \subset (0, t_\sigma) \times K_p$ for all points p sufficiently close to the singularity and some compact set K_p. Whether or not this can be used to infer that these spacetimes are C^0 inextendible remains to be seen.

It should be emphasized that the arguments of [23] are heuristic.

Acknowledgments

We are grateful to the Erwin Schrödinger International Institute for Mathematics and Physics for hospitality during part of work on this project. The research of PTC was supported in part by the Polish National Center of Science (NCN) under grant 2016/21/B/ST1/00940. PK was supported by a uni:docs grant of the University of Vienna. Useful discussions with Greg Galloway are acknowledged.

References

[1] Penrose R 1969 Gravitational collapse: The role of general relativity *Riv. del Nuovo Cim. (numero speziale)* **1** 252–276 URL https://doi.org/10.1023/A:1016578408204

[2] Chruściel P T 1992 On uniqueness in the large of solutions of Einstein equations ("Strong Cosmic Censorship") *Cont. Math.* **132** 235–273

[3] Chruściel P T 1991 *On Uniqueness in the Large of Solutions of Einstein Equations ("Strong Cosmic Censorship")* (Canberra: Australian National University Press)

[4] Sbierski J 2015 The C^0-inextendibility of the Schwarzschild spacetime and the spacelike diameter in Lorentzian geometry arXiv https://arxiv.org/abs/1507.00601

[5] Galloway G J, Ling E and Sbierski J 2017 Timelike completeness as an obstruction to C^0-extensions *Commun. Math. Phys.* ISSN 1432-0916 URL https://doi.org/10.1007/s00220-017-3019-2 arXiv https://arxiv.org/abs/1704.00353

[6] Chruściel P T and Klinger P 2015 Vacuum spacetimes with controlled singularities and without symmetries *Phys. Rev. D* **92** 041501 ISSN 1550-7998 URL http://doi.org/10.1103/PhysRevD.92.041501 arXiv https://arxiv.org/abs/1507.00158

[7] Klinger P 2015 A new class of asymptotically non-chaotic vacuum singularities *Ann. Phys.* **363** 1–35 ISSN 00034916 URL http://doi.org/10.1016/j.aop.2015.09.010 arXiv https://arxiv.org/abs/1507.04161

[8] Isenberg J and Moncrief V 1985 Symmetries of cosmological Cauchy horizons with exceptional orbits *Jour. Math. Phys.* **26** 1024–1027 URL https://doi.org/10.1063/1.526587

[9] Moncrief V and Isenberg J 1983 Symmetries of cosmological Cauchy horizons *Commun. Math. Phys.* **89** 387–413 URL https://projecteuclid.org:443/euclid.cmp/1103922816

[10] Minguzzi E 2015 Area theorem and smoothness of compact Cauchy horizons *Commun. Math. Phys.* **339** 57–98 ISSN 1432-0916 URL https://doi.org/10.1007/s00220-015-2415-8 arXiv https://arxiv.org/abs/1406.5919

[11] Larsson E 2014 Smoothness of compact horizons *Ann. H. Poincaré* 1–52 ISSN 1424-0637 arXiv:1406.6194 [gr-qc] URL http://doi.org/10.1007/s00023-014-0371-z arXiv https://arxiv.org/abs/1406.6194v2

[12] Galloway G J and Ling E 2017 Some remarks on the C^0-(in)extendibility of spacetimes *Ann. H. Poincaré* **18** 3427–3447 ISSN 1424-0661 URL https://doi.org/10.1007/s00023-017-0602-1 arXiv https://arxiv.org/abs/1610.03008

[13] O'Neill B 1983 *Semi-Riemannian Geometry With Applications to Relativity* (*Pure and Applied Mathematics* no 103) (Academic Press)

[14] Sbierski J 2016 private communication

[15] Chruściel P T and Isenberg J 1993 Compact Cauchy horizons and Cauchy surfaces *Directions in General Relativity: Proceedings of the 1993 International Symposium, Maryland: Papers in Honor of Dieter Brill* vol 2 ed Hu B L and Jacobson T A (Cambridge University Press) p 97–107 URL https://doi.org/10.1017/CBO9780511524653.011

[16] Ringström H 2009 Strong cosmic censorship in T^3-Gowdy space-times *Ann. of Math. (2)* **170** 1181–1240 ISSN 0003-486X URL http://doi.org/10.4007/annals.2009.170.1181

[17] Andersson L and Rendall A 2001 Quiescent cosmological singularities *Commun. Math. Phys.* **218** 479–511 ISSN 0010-3616 arXiv https://arxiv.org/abs/gr-qc/0001047

[18] Damour T, Henneaux M, Rendall A D and Weaver M 2002 Kasner-like behaviour for subcritical Einstein-matter systems *Ann. H. Poincaré* **3** 1049–1111 ISSN 1424-0637 URL http://doi.org/10.1007/s000230200000 arXiv https://arxiv.org/abs/gr-qc/0202069

[19] Clausen A and Isenberg J 2007 Areal foliation and asymptotically velocity-term dominated behavior in T^2 symmetric space-times with positive cosmological constant *Jour. Math. Phys.* **48** 1–15 ISSN 00222488 URL http://doi.org/10.1063/1.2767534 arXiv https://arxiv.org/abs/gr-qc/0701054

[20] Ames E, Beyer F, Isenberg J and LeFloch P G 2013 Quasilinear hyperbolic Fuchsian systems and AVTD behavior in T^2-symmetric vacuum spacetimes *Ann. H. Poincaré* **14** 1445–1523 ISSN 1424-0661 URL https://doi.org/10.1007/s00023-012-0228-2 arXiv https://arxiv.org/abs/1205.1881v2

[21] Ringström H 2001 The Bianchi IX attractor *Ann. H. Poincaré* **2** 405–500 ISSN 14240637 URL `http://doi.org/10.1007/PL00001041` arXiv `https://arxiv.org/abs/gr-qc/0006035`

[22] Heinzle J M and Uggla C 2009 Mixmaster: Fact and belief *Class. Quantum Grav.* **26** 1–34 ISSN 0264-9381 URL `http://doi.org/10.1088/0264-9381/26/7/075016` arXiv `https://arxiv.org/abs/0901.0776`

[23] Damour T, Henneaux M and Nicolai H 2003 Cosmological billiards *Class. Quantum Grav.* URL `http://doi.org/10.1088/0264-9381/20/9/201` arXiv `https://arxiv.org/abs/hep-th/0212256`

[24] Heinzle J M, Uggla C and Röhr N 2009 The cosmological billiard attractor *Adv. Theor. Math. Phys.* **13** 293–407 URL `http://projecteuclid.org/euclid.atmp/1234881637` arXiv `https://arxiv.org/abs/gr-qc/0702141`

IOP Conf. Series: Journal of Physics: Conf. Series **968** (2018) 012004 doi:10.1088/1742-6596/968/1/012004

Causal Fermion Systems: A Primer for Lorentzian Geometers

Felix Finster

Fakultät für Mathematik, Universität Regensburg, D-93040 Regensburg, Germany

E-mail: `finster@ur.de`

Abstract. We give a brief introduction to causal fermion systems with a focus on the geometric structures in space-time.

In general relativity, space-time is described by a Lorentzian manifold. Although this description has been highly successful, it is generally believed that in order to reconcile general relativity with quantum theory, the mathematical structure of space-time should be modified on a microscopic scale (typically thought of as the Planck scale $\approx 10^{-35}$ meters). Apart from having far-reaching physical implications, such a modification is also of purely mathematical interest because going beyond the usual space-time continuum makes it possible to model non-smooth structures in space-time and of space-time itself which cannot be described (or at least are difficult to describe) on a Lorentzian manifold.

Causal fermion systems are a recent physical theory based on a novel mathematical model of space-time. The present paper is an introduction to causal fermion systems with a focus on the resulting geometric structures. It is addressed to readers who are familiar with Lorentzian geometry and the Dirac equation. More basic introductions in Minkowski space are given in the textbooks [5, Chapter 1] or [12]. Also, we do not explain the broader physical picture but refer instead to the non-technical survey [10]. Moreover, in [9] some of the concepts are introduced starting from Riemannian geometry. For the physical applications, we refer to [5, Chapters 3-5] and [4]. The relation to other approaches is explained in [5, §3.4.6].

The paper is organized as follows. We first explain how one gets from Lorentzian geometry to the setting of causal fermion systems (Section 1). Then a few structures of an abstract causal fermion system are reviewed (Section 2). In Section 3 we turn attention to the geometric structures and explain how Lorentzian geometry is recovered as a limiting case. In Section 4 it is explained why and specified how the geometric structures of a causal fermion system go beyond Lorentzian geometry. We focus on two aspects: First, space-time need not be smooth and the structures therein need not be regular (Section 4.1). Second, causal fermion systems allow for the description of collections of many different space-times which mutually interact with each other; this is what we mean by "quantum space-time" and "quantum geometry" (Section 4.2). Finally, in Section 5 a few more structures of a causal fermion system are worked out which clarify the connection to other approaches to non-regular space-time geometry (Lorentzian length spaces, causal sets, lattices) or might be useful for the geometric understanding and the future analysis of causal fermion systems.

IOP Conf. Series: Journal of Physics: Conf. Series **968** (2018) 012004 doi:10.1088/1742-6596/968/1/012004

1. From Lorentzian Geometry to Causal Fermion Systems

A basic concept behind causal fermion systems is to encode the geometry in a measure on linear operators on a Hilbert space. We now explain how to get into this setting starting from a usual four-dimensional space-time (for other dimensions see [13, Section 1.4], [14, Section 4] and [9]). More precisely, let $(\mathscr{M}, g)$ be a smooth, globally hyperbolic, time-oriented Lorentzian spin manifold of dimension four. For the signature of the metric we use the convention $(+,-,-,-)$. We denote the corresponding spinor bundle by $S\mathscr{M}$. Its fibers $S_p\mathscr{M}$ are endowed with an inner product $\prec .|. \succ_p$ of signature $(2,2)$. Clifford multiplication is described by a mapping γ which satisfies the anti-commutation relations,

$$\gamma : T_p\mathscr{M} \to \mathrm{L}(S_p\mathscr{M}) \qquad \text{with} \qquad \gamma(u)\,\gamma(v) + \gamma(v)\,\gamma(u) = 2\,g(u,v)\,\mathbb{1}_{S_p(\mathscr{M})}\,. \tag{1}$$

We also write Clifford multiplication in components with the Dirac matrices γ^j. The metric connections on the tangent bundle and the spinor bundle are denoted by ∇. The sections of the spinor bundle are also referred to as wave functions.

We denote the k-times continuously differentiable sections of the spinor bundle by $C^k(\mathscr{M}, S\mathscr{M})$. The Dirac operator $\mathcal{D}$ is defined by

$$\mathcal{D} := i\gamma^j \nabla_j \;:\; C^\infty(\mathscr{M}, S\mathscr{M}) \to C^\infty(\mathscr{M}, S\mathscr{M})\,.$$

Given a real parameter $m \in \mathbb{R}$ (the "mass"), the Dirac equation reads

$$(\mathcal{D} - m)\,\psi = 0\,.$$

We mainly consider solutions in the class $C^\infty_{\text{sc}}(\mathscr{M}, S\mathscr{M})$ of smooth sections with spatially compact support. On such solutions, one has the scalar product

$$(\psi|\phi)_m = 2\pi \int_{\mathscr{N}} \prec \psi \,|\, \gamma(\nu)\,\phi \succ_p d\mu_{\mathscr{N}}(p)\,,$$

where $\mathscr{N}$ denotes any Cauchy surface and ν its future-directed normal (due to current conservation, the scalar product is in fact independent of the choice of $\mathscr{N}$; for details see [14, Section 2]). Forming the completion gives the Hilbert space $(\mathcal{H}_m, (.|.)_m)$.

Next, we choose a closed subspace $\mathcal{H} \subset \mathcal{H}_m$ of the solution space of the Dirac equation. The induced scalar product on $\mathcal{H}$ is denoted by $\langle .|. \rangle_{\mathcal{H}}$. There is the technical difficulty that the wave functions in $\mathcal{H}$ are in general not continuous, making it impossible to evaluate them pointwise. For this reason, we need to introduce an *ultraviolet regularization* on the length scale ε, described mathematically by a linear

$$\textit{regularization operator} \qquad \mathfrak{R}_\varepsilon \;:\; \mathcal{H} \to C^0(\mathscr{M}, S\mathscr{M})\,.$$

In the simplest case, the regularization can be realized by a convolution on a Cauchy surface or in space-time (for details see [14, Section 4] or [5, Section §1.1.2]). For us, the regularization is not just a technical tool, but it realizes the concept already mentioned at the beginning of this paper that we want to change the geometric structures on the microscopic scale. With this in mind, we always consider the regularized quantities as those having mathematical and physical significance. Different choices of regularization operators realize different microscopic space-time structures.

Given $\mathfrak{R}_\varepsilon$, for any space-time point $p \in \mathscr{M}$ we consider the bilinear form

$$b_p \;:\; \mathcal{H} \times \mathcal{H} \to \mathbb{C}\,, \quad b_p(\psi, \phi) = -\prec (\mathfrak{R}_\varepsilon \psi)(p) \,|\, (\mathfrak{R}_\varepsilon \phi)(p) \succ_p\,.$$

This bilinear form is well-defined and bounded because $\mathfrak{R}_\varepsilon$ maps to the continuous wave functions and because evaluation at p gives a linear operator of finite rank. Thus for any $\phi \in \mathcal{H}$, the anti-linear form $b_p(.,\phi) : \mathcal{H} \to \mathbb{C}$ is continuous. By the Fréchet-Riesz theorem, there is a unique $\chi \in \mathcal{H}$ such that $b_p(\psi,\phi) = \langle \psi | \chi \rangle_\mathcal{H}$ for all $\psi \in \mathcal{H}$. The mapping $\phi \mapsto \chi$ is linear and bounded. We thus obtain a unique bounded linear operator $F^\varepsilon(p)$ on $\mathcal{H}$ which is characterized by the relation

$$(\psi \,|\, F^\varepsilon(p)\, \phi) = -\prec(\mathfrak{R}_\varepsilon \psi)(p) | (\mathfrak{R}_\varepsilon \phi)(p)\succ_p \qquad \text{for all } \psi, \phi \in \mathcal{H}\,. \tag{2}$$

Taking into account that the inner product on the Dirac spinors at p has signature $(2,2)$, the local correlation operator $F^\varepsilon(p)$ is a symmetric operator on $\mathcal{H}$ of rank at most four, which (counting multiplicities) has at most two positive and at most two negative eigenvalues. Varying the space-time point, we obtain a mapping

$$F^\varepsilon \,:\, \mathscr{M} \to \mathcal{F} \subset \mathrm{L}(\mathcal{H})\,,$$

where $\mathcal{F}$ denotes all symmetric operators of rank at most four with at most two positive and at most two negative eigenvalues. Finally, we introduce the

$$\textit{universal measure} \qquad d\rho := (F^\varepsilon)_* \, d\mu_\mathscr{M}$$

as the push-forward of the volume measure on $\mathscr{M}$ under the mapping F^ε (thus $\rho(\Omega) := \mu_\mathscr{M}((F^\varepsilon)^{-1}(\Omega))$).

In this way, we obtain a measure ρ on the set $\mathcal{F} \subset \mathrm{L}(\mathcal{H})$ of linear operators on a Hilbert space $\mathcal{H}$. The basic concept is to work exclusively with these objects, but to drop all other structures (like the Lorentzian metric g, the structure of the spinor bundle $S\mathscr{M}$, the manifold structure of $\mathscr{M}$, and even the structure of $\mathscr{M}$ being a point set). This leads us to the structure of a causal fermion system of spin dimension two, as will be defined abstractly at the beginning of the next section.

Before turning attention to the abstract setting, we make a few comments on the underlying physical picture. The vectors in the subspace $\mathcal{H} \subset \mathcal{H}_m$ have the interpretation as those Dirac wave functions which are realized in the physical system under consideration. If we describe for example a system of one electron, then the wave function of the electron is contained in $\mathcal{H}$. Moreover, $\mathcal{H}$ includes all the wave functions which form the so-called Dirac sea (for an explanation of this point see for example [3]). We refer to the vectors in $\mathcal{H}$ as the *physical wave functions*. The name causal *fermion* system is motivated by the fact that Dirac particles are fermions. According to (2), the local correlation operator $F^\varepsilon(p)$ describes densities and correlations of the physical wave functions at the space-time point p. Working exclusively with the local correlation operators and the corresponding push-forward measure ρ means in particular that the geometric structures are encoded in and must be retrieved from the physical wave functions. Since the physical wave functions describe the distribution of matter in space-time, one can summarize this concept by saying that *matter encodes geometry.*

2. The Abstract Setting and a Few Inherent Structures

2.1. Basic Definition of a Causal Fermion System

We now give the abstract definition of a causal fermion system:

Definition 2.1 Given a separable complex Hilbert space $\mathcal{H}$ with scalar product $\langle .|. \rangle_\mathcal{H}$ and a parameter $n \in \mathbb{N}$ (the *"spin dimension"*), we let $\mathcal{F} \subset \mathrm{L}(\mathcal{H})$ be the set of all self-adjoint operators on $\mathcal{H}$ of finite rank, which (counting multiplicities) have at most n positive and at most n negative eigenvalues. On $\mathcal{F}$ we are given a measure ρ (defined on a σ-algebra of subsets of $\mathcal{F}$), the so-called *universal measure.* We refer to $(\mathcal{H}, \mathcal{F}, \rho)$ as a *causal fermion system.*

IOP Conf. Series: Journal of Physics: Conf. Series **968** (2018) 012004 doi:10.1088/1742-6596/968/1/012004

This definition, which was first given in [8, Section 1.2], no longer involves the usual geometric structures (like a manifold $\mathscr{M}$, a Lorentzian metric g or a spinor bundle $S\mathscr{M}$). Therefore, it is far from obvious why and how a causal fermion system should provide a setting for geometry in space-time. All we can say right away is that a causal fermion system encodes a lot of information. For example, parts of this information can be retrieved by taking traces of products of operators in $\mathcal{F}$ and integrating,

$$\int_{\mathcal{F}} \text{tr}(x)\, d\rho(x)\,, \qquad \int_{\mathcal{F}} d\rho(x) \int_{\mathcal{F}} d\rho(y)\; \text{tr}(x \cdot y)\,, \qquad \ldots\,.$$

The point is that by describing all this information with useful and convenient notions, one recovers a space-time as well as geometric structures therein. All these structures are *inherent* in the sense that they are no additional input, but they only give information which is already encoded in the causal fermion system an apposite name, thereby providing a better intuitive understanding of the causal fermion system.

Working with the inherent structures, causal fermion systems provide a general mathematical framework in which there are many analytic, geometric and topological structures. In particular, it becomes possible to generalize notions of differential geometry to the non-smooth setting. From the physical point of view, causal fermion systems are a proposal for quantum geometry and an approach to quantum gravity. Giving quantum mechanics, general relativity and quantum field theory as limiting cases, they are a candidate for a unified physical theory. For the physical applications, the key point is that the physical equations can be formulated within the setting of causal fermion systems in terms of a variational principle called the *causal action principle* (see [5, Section 1.1]). The causal action principle is the analytic core of the theory. For brevity we cannot introduce the causal action principle here, but instead refer the interested reader to the mathematical introduction in the recent paper [11] and to the references therein.

2.2. Space-Time and Causal Structure

We now introduce a few inherent structures of a causal fermion system $(\mathcal{H}, \mathcal{F}, \rho)$. *Space-time* M is defined as the support of the universal measure[1],

$$M := \text{supp}\, \rho \subset \mathcal{F}\,. \tag{3}$$

On M we consider the topology induced by $\mathcal{F}$ (generated by the sup-norm on $\text{L}(\mathcal{H})$). Typically, M is a low-dimensional subset of $\mathcal{F}$ which can have smooth but also discrete or non-regular components (see Figure 1).

The next definition is the reason for the name *causal* fermion system.

Definition 2.2 *(causal structure)* For any $x, y \in M$, the product xy is an operator of rank at most $2n$. We denote its non-trivial eigenvalues (counting algebraic multiplicities) by $\lambda^{xy}_1, \ldots, \lambda^{xy}_{2n}$. The points x and y are called *spacelike* separated if all the λ^{xy}_j have the same absolute value. They are said to be *timelike* separated if the λ^{xy}_j are all real and do not all have the same absolute value. In all other cases (i.e. if the λ^{xy}_j are not all real and do not all have the same absolute value), the points x and y are said to be *lightlike* separated.

We remark that this definition is closely linked to the structure of the causal action (see [5, §1.1.2]).

[1] The *support* of a measure is defined as the complement of the largest open set of measure zero, i.e.

$$\text{supp}\, \rho := \mathcal{F} \setminus \bigcup \big\{ \Omega \subset \mathcal{F} \;\big|\; \Omega \text{ is open and } \rho(\Omega) = 0 \big\}\,.$$

It is by definition a closed set.

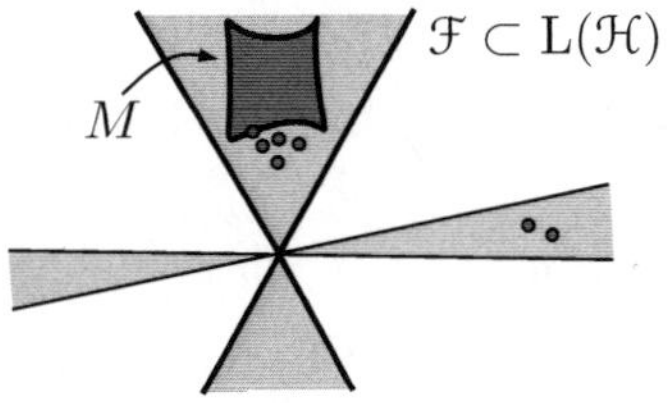

Figure 1. Space-time M of a causal fermion system

A causal fermion system also distinguishes a *direction of time*. To this end, we let π_x be the orthogonal projection in $\mathcal{H}$ on the subspace $x(\mathcal{H}) \subset \mathcal{H}$ and introduce the functional

$$\mathcal{C} \,:\, M \times M \to \mathbb{R}\,, \qquad \mathcal{C}(x,y) := i \,\text{tr}\big(y\,x\,\pi_y\,\pi_x - x\,y\,\pi_x\,\pi_y\big) \tag{4}$$

(this functional was first stated in [9, Section 8.5], motivated by constructions in [7, Section 3.5]). Obviously, this functional is anti-symmetric in its two arguments. This makes it possible to introduce the notions

$$\left\{ \begin{array}{ll} y \text{ lies in the } \textit{future} \text{ of } x & \text{if } \mathcal{C}(x,y) > 0 \\ y \text{ lies in the } \textit{past} \text{ of } x & \text{if } \mathcal{C}(x,y) < 0\,. \end{array} \right. \tag{5}$$

We remark that in suitable limiting cases, the above notions of causality indeed agree with the causal structure of a Lorentzian space-time (for details see [5, §1.2.5] or [7, Sections 4 and 5]).

2.3. Spinors and Physical Wave Functions
For every $x \in \mathcal{F}$ we define the *spin space* S_xM by $S_xM = x(\mathcal{H})$; it is a subspace of $\mathcal{H}$ of dimension at most $2n$. On the spin space S_xM the *spin scalar product* $\prec .|.\succ_x$ is defined by

$$\prec .|.\succ_x : S_xM \times S_xM \to \mathbb{C}\,, \qquad \prec u|v\succ_x = -\langle u|xu\rangle_{\mathcal{H}}\,, \tag{6}$$

making the spin space $(S_xM, \prec .|.\succ_x)$ to an indefinite inner product of signature (p,q) with $p, q \leq n$.

A *wave function* ψ is defined as a function which to every $x \in M$ associates a vector of the corresponding spin space,

$$\psi \,:\, M \to \mathcal{H} \qquad \text{with} \qquad \psi(x) \in S_xM \quad \text{for all } x \in M\,. \tag{7}$$

To every vector $u \in \mathcal{H}$ we can associate a wave function ψ^u obtained by taking the orthogonal projections to the corresponding spin spaces,

$$\psi^u \,:\, M \to \mathcal{H}\,, \qquad \psi^u(x) = \pi_x u \in S_xM$$

(with π_x as defined before (4)). We refer to ψ^u as the *physical wave function* of $u \in \mathcal{H}$ (we remark that, for the causal fermion systems obtained by the construction in Section 1, after suitable identifications made, the physical wave function ψ^u indeed coincides with the regularized Dirac wave function $\mathfrak{R}_\varepsilon u$; for details see [5, §1.2.4]).

In this way, every vector of $\mathcal{H}$ can be represented by a wave function in space-time. The resulting ensemble of all physical wave functions is described most conveniently by the *kernel of the fermionic projector* $P(x,y)$ defined by

$$P(x,y) = \pi_x\, y|_{S_yM} \,:\, S_yM \to S_xM\,. \tag{8}$$

Due to the factor x in the definition of the spin scalar product (6), the kernel of the fermionic projector is symmetric in the sense that $P(x,y)^* = P(y,x)$, as is immediately verified by the computation

$$\begin{aligned} \prec u \,|\, P(x,y)\, v \succ_x &= -\langle u \,|\, x\, P(x,y)\, v\rangle_{\mathcal{H}} = -\langle u \,|\, xy\, v\rangle_{\mathcal{H}} \\ &= -\langle \pi_y\, x\, u \,|\, y\, v\rangle_{\mathcal{H}} = \prec P(y,x)\, u \,|\, v \succ_y \qquad (\text{for } u \in S_xM,\ v \in S_yM)\,. \end{aligned}$$

3. Geometric Structures of a Causal Fermion System

In [7] the geometric structures of a causal fermion system were worked out abstractly, and it was shown that in a certain limiting case one recovers Lorentzian spin geometry. We now outline a few constructions and results from this paper.

3.1. Regular Causal Fermion Systems

According to Definition 2.1, the operators in $M \subset \mathcal{F}$ have at most n positive and at most n negative eigenvalues. In most situations of physical interest and most examples, the number of positive and negative eigenvalues equals n. This motivates the following definition:

Definition 3.1 A space-time point $x \in M$ is said to be *regular* if x has the maximal possible rank, i.e. $\dim x(\mathcal{H}) = 2n$. Otherwise, the space-time point is called *singular*. A causal fermion system is *regular* if all its space-time points are regular.

For a regular causal fermion system, the set

$$SM := \bigcup_{x \in M} S_xM$$

has the structure of a topological vector bundle with base M and fibers S_xM (for details and all topological issues see [9]). The fibers are endowed with an inner product $\prec .|. \succ_p$ of signature (n,n). A wave function (7) is a section of this bundle (for the notion of continuity of such wave functions and the space $C^0(M, SM)$ we refer to [5, §1.1.4]). As a specific feature, all the fibers are subspaces of the same Hilbert space $(\mathcal{H}, \langle .|. \rangle_{\mathcal{H}})$.

3.2. Clifford Subspaces

The structure of wave functions (7) taking values in the spin spaces is reminiscent of sections of a vector bundle. However, one important structure is missing: we have no Dirac matrices and no notion of Clifford multiplication. The following definition is a step towards introducing these additional structures.

Definition 3.2 *(Clifford subspace)* We denote the space of symmetric linear operators on $(S_xM, \prec .|. \succ_x)$ by $\text{Symm}(S_xM) \subset \text{L}(S_xM)$. A subspace $K \subset \text{Symm}(S_xM)$ is called a *Clifford subspace* of signature (r,s) at the point x (with $r, s \in \mathbb{N}_0$) if the following conditions hold:

(i) For any $u, v \in K$, the anti-commutator $\{u, v\} \equiv uv + vu$ is a multiple of the identity on S_xM.

(ii) The bilinear form $\langle .,. \rangle$ on K defined by

$$\frac{1}{2}\,\{u,v\} = \langle u,v \rangle\, \mathbb{1} \qquad \text{for all } u, v \in K \tag{9}$$

is non-degenerate and has signature (r,s).

In view of the anti-commutation relations (9), a Clifford subspace can be regarded as a generalization of the space spanned by the usual Dirac matrices. However, the above definition has two shortcomings: First, there are many different Clifford subspaces, so that there is no unique notion of Clifford multiplication. Second, we are missing the structure of tangent vectors. We now explain how to overcome these shortcomings.

IOP Conf. Series: Journal of Physics: Conf. Series **968** (2018) 012004 doi:10.1088/1742-6596/968/1/012004

3.3. Connection and Curvature

The kernel of the kernel of the fermionic projector (8) is a mapping from one spin space to another and thus induces relations between space-time points. The idea is to use these relations for the construction of a spin connection $D_{x,y}$, being a unitary mapping between the corresponding spin spaces,

$$D_{x,y} \;:\; S_yM \rightarrow S_xM$$

(we consistently use the notation that the subscript $_{xy}$ denotes an object at the point x, whereas the additional comma $_{x,y}$ denotes an operator which maps an object at y to an object at x). The simplest idea for the construction of the spin connection would be to form a polar decomposition $P(x,y) = SU$ with a symmetric operator $S : S_xM \rightarrow S_xM$ and a unitary operator $U : S_yM \rightarrow S_xM$ and to introduce the spin connection as the unitary part, $D_{x,y} = U$. However, this method is too naive, because we want the spin connection to be compatible with a corresponding metric connection $\nabla_{x,y}$ which should map Clifford subspaces at x and y (see Definition 3.2 above) isometrically to each other. Another complication is that, as explained after Definition 3.2, the Clifford subspaces at x and y are not unique. The method to bypass these problems is to work with several Clifford subspaces and to use so-called splice maps, as we now briefly explain.

First, it is useful to restrict the freedom in choosing the Clifford subspaces with the following construction. Recall that for any $x \in M$, the operator $(-x)$ on $\mathcal{H}$ has at most n positive and at most n negative eigenvalues. We denote its positive and negative spectral subspaces by S_x^+M and S_x^-M, respectively. In view of (6), these subspaces are also orthogonal with respect to the spin scalar product,

$$S_xM = S_x^+M \oplus S_x^-M\,.$$

We introduce the *Euclidean sign operator* s_x as a symmetric operator on S_xM whose eigenspaces corresponding to the eigenvalues ± 1 are the spaces S_x^+M and S_x^-M, respectively. Since $s_x^2 = \mathbb{1}$, the span of the Euclidean sign operator is a one-dimensional Clifford subspace of signature $(1,0)$. The idea is to extend s_x to obtain higher-dimensional Clifford subspaces. We thus define a *Clifford extension* as a Clifford subspace which contains s_x. By restricting attention to Clifford extensions, we have reduced the freedom in choosing Clifford subspaces. However, still there is not a unique Clifford extension, even for fixed dimension and signature. But one can define the *tangent space* T_x as an equivalence class of Clifford extensions; for details see [7, Section 3.1]. Choosing $r = 1$, the bilinear form $\langle .,. \rangle$ in (9) induces a Lorentzian metric on the tangent space.

In the remainder of this section, we assume that the causal fermion system is regular (see Definition 3.1). Moreover, we need the following stronger version of timelike separation:

Definition 3.3 The space-time point $x \in M$ is *properly timelike* separated from $y \in M$ if the closed chain A_{xy} defined by

$$A_{xy} = P(x,y)\,P(y,x) \;:\; S_xM \rightarrow S_xM$$

has a strictly positive spectrum and if all eigenspaces are (either positive or negative) definite subspaces of $(S_xM, \prec .|.\succ_x)$.

The two following observations explain why the last definition makes sense (for the proof see [5, §1.1.6]):

- Properly timelike separation implies timelike separation (see Definition 2.2).
- The notion is symmetric in x and y.

So far, the construction of the spin connection has been worked out only in the case of spin dimension $n = 2$. Then for two properly timelike separated points $x, y \in M$, the spin space S_xM can be decomposed uniquely into an orthogonal direct sum $S_xM = I^+ \oplus I^-$ of a two-dimensional positive definite subspace I^+ and a two-dimensional negative definite subspace I^- of A_{xy}. We define the *directional sign operator* v_{xy} of A_{xy} as the unique operator on S_xM such that the eigenspaces corresponding to the eigenvalues ± 1 are the subspaces $I^\pm$.

Having the Euclidean sign operator s_x and the directional sign operator v_{xy} to our disposal, under generic assumptions one can distinguish two Clifford subspaces at the point x: a Clifford subspace K_{xy} containing v_{xy} and a Clifford extension $K_x^{(y)}$ (for details see [7, Lemma 3.12]). Similarly, at the point y we have a distinguished Clifford subspace K_{yx} (which contains v_{yx}) and a distinguished Clifford extension $K_y^{(x)}$. For the construction of the *spin connection* $D_{x,y} : S_yM \to S_xM$ one works with the Clifford subspaces K_{xy} and K_{yx} and demands that these are mapped to each other. More precisely, the spin connection is uniquely characterized by the following properties (see [7, Theorem 3.20]):

(i) $D_{x,y}$ is of the form

$$D_{x,y} = e^{i\varphi_{xy} v_{xy}} A_{xy}^{-\frac{1}{2}} P(x,y) \quad \text{with} \quad \varphi_{xy} \in \Big(-\frac{3\pi}{4}, -\frac{\pi}{2}\Big) \cup \Big(\frac{\pi}{2}, \frac{3\pi}{4}\Big) .$$

(ii) The spin connection maps the Clifford subspaces K_{xy} and K_{yx} to each other, i.e.

$$D_{y,x} K_{xy} D_{x,y} = K_{yx} .$$

The spin connection has the properties

$$D_{y,x} = (D_{x,y})^{-1} = (D_{x,y})^* \qquad \text{and} \qquad A_{xy} = D_{x,y} A_{yx} D_{y,x} .$$

All the assumptions needed for the construction of the spin connection are combined in the notion that x and y must be *spin-connectable* (see [7, Definition 3.17]).

By composing the spin connection along a "discrete path" of space-time points, one obtains a "parallel transport" of spinors. When doing so, it is important to keep track of the different Clifford subspaces and to carefully transform them to each other. In order to illustrate in an example how this works, suppose that we want to compose the spin connection $D_{y,z}$ with $D_{z,x}$. As mentioned above, the spin connection $D_{z,x}$ at the point z is constructed using the Clifford subspace K_{zx}. The spin connection $D_{y,z}$, however, takes at the same space-time point z the Clifford subspace K_{zy} as reference. This entails that before applying $D_{y,z}$ we must transform from the Clifford subspace K_{zx} to the Clifford subspace K_{zy}. This is accomplished by the *splice map* $U_z^{(y|x)}$, being a uniquely defined unitary transformation of S_xM with the property that

$$K_{zy} = U_z^{(y|x)} K_{zx} \big(U_z^{(y|x)}\big)^* .$$

The splice map must be sandwiched between the spin connections in combinations like

$$D_{y,z} U_z^{(y|x)} D_{z,x} .$$

In order to construct a corresponding metric connection $\nabla_{x,y}$, one uses a similar procedure to relate the Clifford subspaces to corresponding Clifford extensions. More precisely, one first unitarily transforms the Clifford extension $K_y^{(x)}$ to the Clifford subspace K_{yx}. Unitarily transforming with the spin connection D_{xy} gives the Clifford subspace K_{xy}. Finally, one unitarily transforms to the Clifford extension $K_x^{(y)}$. Since the Clifford extensions at the beginning and end are representatives of the corresponding tangent spaces, we thus obtain an isometry

$$\nabla_{x,y} : T_y \to T_x$$

between the tangent spaces (for details see [7, Section 3.4]).

In this setting, *curvature* is defined as usual as the holonomy of the connection. Thus the curvature of the spin connection is given by

$$\mathfrak{R}(x,y,z) = U_x^{(z|y)} D_{x,y} U_y^{(x|z)} D_{y,z} U_z^{(y|x)} D_{z,x} : S_xM \to S_xM ,$$

and similarly for the metric connection.

3.4. Correspondence to Lorentzian Geometry

In [7, Sections 4 and 5] it is proven that the above notions reduce to the metric connections and the Riemannian curvature on a globally hyperbolic Lorentzian manifold if the causal fermion system is constructed by regularizing solutions of the Dirac equation (similar as explained in Section 1) and removing the regularization in a suitable way. These results show that the notions of connection and curvature defined above indeed generalize the corresponding notions in Lorentzian spin geometry.

As in Section 1, we begin with a globally hyperbolic space-time $(\mathscr{M}, g)$. In order to keep the setting as simple as possible, in [7] it was assumed that $(\mathscr{M}, g)$ is isometric to Minkowski space in the past of a Cauchy-hypersurface $\mathscr{N}$. This has the technical advantage that one can work with simple and explicit regularization operators $\mathcal{R}_\varepsilon$ defined in Minkowski space (more precisely, one works with $i\varepsilon$-regularization of a Dirac see structure; see also [5, §2.4.1]). Due to the ultraviolet regularization, the causal fermion system only describes the "coarse geometry" down to the length scale ε. Therefore, in order to recover Lorentzian spin geometry, we need to analyze the limit $\varepsilon \searrow 0$.

For sufficiently small ε, one can identify objects of Lorentzian spin geometry with corresponding objects of the causal fermion system in the following way:

Lorentzian geometry	**causal fermion system**
space-time manifold $\mathscr{M}$	space-time $M := \operatorname{supp} \rho \subset \mathcal{F} \subset \mathrm{L}(\mathcal{H})$
space-time point $p \in \mathscr{M}$	space-time point $x := F^\varepsilon(p) \in M$
spinor space $S_p\mathscr{M}$	spin space $S_xM := x(\mathcal{H}) \subset \mathcal{H}$
wave function $u \in \mathcal{H} \subset \mathcal{H}_m$	physical wave function $\psi^u : M \to \mathcal{H}$
spinor $u(p) \in S_p\mathscr{M}$	spinor $\psi^u(x) \in S_xM$

Assume that $\gamma : [0, T] \to \mathscr{M}$ is a future-directed timelike curve joining the points $q := \gamma(0)$ with $p := \gamma(T)$, for simplicity parametrized by arc length. In order to avoid confusion with the different connections, we now denote the metric connection on the tangent bundle $T\mathscr{M}$ (i.e. the Levi-Civita connection) by ∇^{LC}. Likewise, the metric connection on the spinor bundle $S\mathscr{M}$ is denoted by D^{LC}.

In order to compare the parallel transports $D^{\text{LC}}_{p,q}$ and ∇^{LC} along γ with the corresponding connections D and ∇ of the causal fermion system, given $N \in \mathbb{N}$ we choose intermediate points on the curve by

$$p_n := \gamma\Big(\frac{nT}{N}\Big) , \qquad n = 0, \ldots, N . \tag{10}$$

The following results show that in the limit $\varepsilon \searrow 0$ and $N \to \infty$, the connections agree, up to higher orders in the curvature tensor (for the proof see [7, Theorem 5.12 and Corollary 5.13]).

Theorem 3.4 *There is a subset of curves which is dense in the C^∞-topology with the following property: Choosing $N \in \mathbb{N}$ sufficiently large and $\varepsilon > 0$ sufficiently small, the points p_n and p_{n-1} are spin-connectable for all $n = 1, \ldots, N$, and every point p_n lies in the future of p_{n-1} (in the* *sense* (5)*). Moreover,*

$$\begin{aligned}
&\lim_{N\to\infty} \lim_{\varepsilon\searrow 0} D_{(p_N,p_{N-1})}\, D_{(p_{N-1},p_{N-2})} \cdots D_{(p_1,p_0)} \\
&\quad = D^{\text{LC}}_{p,q}\, \mathrm{Texp}\bigg(\frac{1}{6} \int_\gamma \Big(m^2 - \frac{\mathrm{scal}}{12}\Big)^{-1} \\
&\qquad\qquad \times D^{\text{LC}}_{q,\gamma(t)} \Big[\epsilon_j \,(\nabla_{e_j} R)\big(\dot{\gamma}(t), e_j\big)\, \dot{\gamma}(t) \Big] \cdot \dot{\gamma}(t) \cdot D^{\text{LC}}_{\gamma(t),q}\, dt \bigg) \\
&\lim_{N\to\infty} \lim_{\varepsilon\searrow 0} \nabla_{p_N,p_{N-1}}\, \nabla_{p_{N-1},p_{N-2}} \cdots \nabla_{p_1,p_0} \\
&\quad = \nabla^{\text{LC}}_{q,p} + \mathcal{O}\bigg(L(\gamma)\, \frac{\|\nabla R\|}{m^2} \bigg) \bigg(1 + \mathcal{O}\Big(\frac{\mathrm{scal}}{m^2} \Big) \bigg) ,
\end{aligned}$$

where Texp *is the time-ordered exponential, and* $L(\gamma)$ *is the length of the curve* γ*.*

4. Going Beyond Lorentzian Geometry

4.1. Non-Smooth Geometries

The construction of connection and curvature in Section 3.3 did no rely on a smooth space-time structure. Indeed, these constructions even apply in non-regular or discrete space-times (for many examples see [9, Section 9]). Therefore, causal fermion systems are a framework for non-smooth geometries. We now explain in words how this works and what to keep in mind. Suppose that a causal fermion system $(\mathcal{H}, \mathcal{F}, \rho)$ is given. If space-time $M := \operatorname{supp} \rho$ has the structure of a smooth manifold, then we are in the smooth setting. Otherwise, M merely is a topological space, with additional structures induced by the fact that the space-time points are linear operators in $\mathcal{F}$. The singular points (see Definition 3.1) must be treated separately. Removing them from M (by multiplying ρ with the characteristic function of the regular points), one gets a regular causal fermion system. Then the constructions in Sections 3.2 and 3.3 apply, giving a spin connection $D_{x,y}$, a metric connection $\nabla_{x,y}$ as well as corresponding curvatures.

A subtle point to keep in mind is that, since M has no manifold structure, there is no notion of tangent vectors. This also means that there is no analog of Clifford multiplication (1). Instead, the Clifford subspace (see Definition 3.2) merely is a vector space of linear operators acting on the spin space endowed with an inner product $\langle .,. \rangle$, but the vectors in the Clifford subspace are not related to usual tangent vectors defined as equivalence classes of curves or as derivations.

One method for getting along without tangent vectors is to work with *tangent cone measures*, as we now briefly outline (another method, which will be explained in Section 5.1 below, is to construct a corresponding Lorentzian length space). In the simplest version of this construction, one first chooses a mapping $\mathcal{A}$ from space-time to the vector space $\mathrm{Symm}(S_x)$ of all symmetric linear operators on the spin space,

$$\mathcal{A} : M \to \mathrm{Symm}(S_x) \qquad \text{with} \qquad \mathcal{A}(u) = \pi_x\,(y - x)\, x|_{S_x} \,. \tag{11}$$

Then the push-forward $\mathcal{A}_*(\rho)$ is a measure on $\mathrm{Symm}(S_x)$. A *conical set* is a subset of $\mathrm{Symm}(S_x)$ with $\mathbb{R}^+ A = A$. The intuitive idea is to define the tangent cone measure μ_{con} at the space-time point x as a measure on conical sets obtained by restriction to smaller and smaller neighborhoods of x and rescaling, i.e.

$$\mu_{\text{con}}(A) := \liminf_{\delta\searrow 0} \frac{1}{\rho\big(B_\delta(x)\big)}\, \rho\Big(\mathcal{A}^{-1}(A_k) \cap B_\delta(x) \Big)$$

(where $B_\delta(x) \subset \mathrm{L}(\mathcal{H})$ is the Banach space ball). In order to get σ-additivity, this definition must be modified and complemented by measure-theoretic constructions (for details see [9, Section 6]).

IOP Conf. Series: Journal of Physics: Conf. Series **968** (2018) 012004 doi:10.1088/1742-6596/968/1/012004

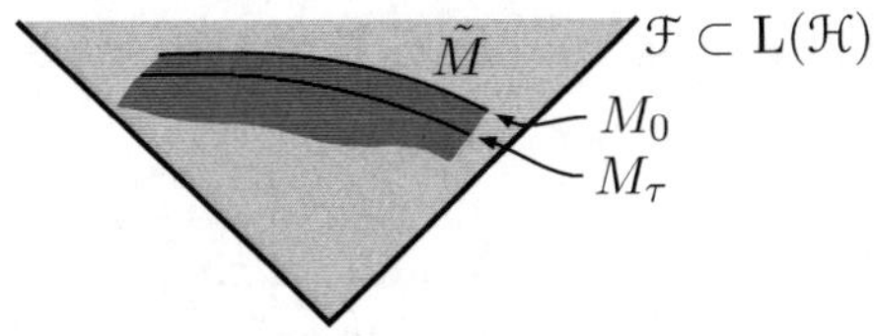

Figure 2. A quantum space-time

Such tangent cone measures make it possible to analyze the local structure of space-time in a neighborhood of a point $x \in M$, again without any differentiability assumptions. In particular, the tangent cone measures can be used to distinguish a specific Clifford subspace $\mathcal{C}\ell_x$ and to relate $\mathcal{C}\ell_x$ to neighboring space-time points (for details see again [9, Section 6]).

4.2. Quantum Geometries

The general definition of causal fermion systems (see Definition 2.1) covers many situation in which space-time $M := \operatorname{supp} \rho$ does not have the structure of a manifold or a space-time lattice and cannot be thought of as a classical space-time. Since in our setting the geometric structure are encoded in the quantum matter, we subsume the general situation under the notion *quantum geometry.* This name is also motivated by the fact that these more general space-times are indeed relevant for the applications to quantum field theory (see the mechanism of *microscopic mixing* in [4] or the related *fragmentation of space-times* in [6]).

We now illustrate in a simple example what a "quantum space-time" is about. To this end, we generalize the setting of Section 1 by considering on $\mathscr{M}$ a family of Lorentzian metrics g_τ indexed by a parameter $\tau \in [0, 1]$ (for example, the family can be obtained by varying the metric inside a compact subset $K \subset \mathscr{M}$). We assume that for every τ, the manifold $(\mathscr{M}, g_\tau)$ is globally hyperbolic, time-oriented and spin. Moreover, we identify the Hilbert spaces $\mathcal{H}_m$ for different values of τ (for example by identifying boundary values on a Cauchy surface which does not intersect K). Then the construction of Section 1 gives for every $\tau \in [0, 1]$ a measure ρ_τ on $\mathcal{F}$. The equation

$$\tilde{\rho}(\Omega) = \int_0^1 \rho_\tau(\Omega)\, d\tau$$

(where $d\tau$ is the Lebesgue measure) defines a measure on $\mathcal{F}$. The resulting space-time $\tilde{M}$ as given by

$$\tilde{M} := \operatorname{supp} \tilde{\rho} = \overline{\bigcup_{\tau \in (0,1)} M_\tau} \qquad \text{with} \qquad M_\tau := \operatorname{supp} \rho_\tau$$

can no longer be identified with the original space-time manifold $\mathscr{M}$, but it can be thought of as a collection of the whole family of space-times $(\mathscr{M}, g_\tau)_{\tau \in [0,1]}$.

In other words, the space-time $\tilde{M}$ is no longer described by a single metric, but by the whole family of metrics (g_τ). The situation becomes even more interesting if an interaction (as described by the causal action principle) is taken into account. Then the different "copies" M_τ of space-time interact with each other (as made precise by the constructions in [6]), giving rise to a complicated measure which can no longer be described in terms of a family of classical metrics. But, as shown in [6], the resulting measure can be described in terms of quantum fields. This justifies the name "quantum space time."

IOP Conf. Series: Journal of Physics: Conf. Series **968** (2018) 012004 doi:10.1088/1742-6596/968/1/012004

5. Connection to Other Approaches and Further Structures

5.1. Connection to Lorentzian Length Spaces

A recent approach to non-smooth Lorentzian geometry is provided by Lorentzian length spaces [15]. We now explain how a causal fermion system encodes the structure of a Lorentzian length space. Our construction involves the choice of two length scales $\ell_{\min}$ and $\ell_{\max}$. This is a subtle point which we explain in detail. To this end, we return to the setting of a globally hyperbolic space-time $(\mathscr{M}, g)$ and consider a "discrete timelike path" (10) of space-time points $p_0, \ldots, p_N \in \mathscr{M}$, such that the Lorentzian distance of adjacent points is of the order $\ell > 0$. In order to get a good approximation of a continuous path, the points should be sufficiently close to each other. The relevant length scale $\ell_{\max}$ must be chosen so small that curvature effects are not relevant on this scale. Moreover, we choose $\ell_{\max}$ so small that the oscillations of the Dirac wave functions due to the rest mass do not come into play. Thus we choose $\ell_{\max}$ to be smaller than the Compton scale m^{-1}. The length scale $\ell_{\min}$, on the other hand, must be chosen so large that the effects of the regularization on the scale ε do not yet come into play (where ε can again be thought of as the Planck length). We thus obtain the admissible range

$$\varepsilon \ll \ell_{\min} < \ell < \ll \ell_{\max} \ll m^{-1}\,, \tag{12}$$

leaving us with a lot of freedom to choose the parameters $\ell_{\min}$ and $\ell_{\max}$ (note that for electrons, $m\varepsilon \lesssim 10^{-23}$). If the Lorentzian geodesic distance ℓ is within the range (12), then the kernel of the fermionic projector of the corresponding causal fermion system is well-approximated by the unregularized massless kernel in Minkowski space (as computed in detail in [5, Section 1.2]). In particular, the eigenvalues of xy scale like ℓ^{-6}. Since these eigenvalues are defined in terms of objects of the corresponding causal fermion system, we can generalize the notion the Lorentzian distance as follows.

Given a causal fermion system $(\mathcal{H}, \mathcal{F}, \rho)$, we introduce the functional

$$\ell \,:\, M \times M \to \mathbb{R}_0^+\,, \qquad \ell(x,y) = \begin{cases} |xy|^{-\frac{1}{6}} & \text{if } \ell_{\min} < |xy|^{-\frac{1}{6}} < \ell_{\max} \\ 0 & \text{otherwise}\,. \end{cases}$$

A finite sequence of points $(x_0, \ldots, x_N) \in M$ is called a *causal chain* if for all $n = 1, \ldots, N$ the following conditions hold:

(i) x_n and x_{n-1} are timelike separated
(ii) x_n lies in the future of x_{n-1}
(iii) $\ell(x_n, x_{n+1}) > 0$

The condition (i) could be strengthened by demanding that x_n and x_{n+1} are spin-connectable (see [7, Definition 3.17]). The causal chain is said to *connect* the space-time point x with y if $x_0 = x$ and $x_N = y$. The *length* of the causal chain is defined by

$$L[x_0, \ldots, x_N] = \sum_{n=1}^N \ell\big(x_{n-1}, x_n\big)\,.$$

The *Lorentzian distance* of two points is defined by

$$\begin{aligned} d(x,y) = \sup \big\{ L[x_0, \ldots, x_N] \;\big|\; \\ (x_0, \ldots, x_N) \text{ is a causal chain connecting } x \text{ with } y \big\}\,. \end{aligned} \tag{13}$$

If the set on the right is empty, then the supremum is defined to be zero. We also point out that the last supremum could be infinite, even if the points x and y are close together. A typical

IOP Conf. Series: Journal of Physics: Conf. Series **968** (2018) 012004 doi:10.1088/1742-6596/968/1/012004

example when this happens is when there are closed causal chains (corresponding to closed timelike curves on a Lorentzian manifold).

With the above definitions, a causal fermion system indeed gives rise to a Lorentzian length space (our setting agrees precisely with [15], up to a few additional technical assumptions in [15] which would have to be verified in the applications).

5.2. Connection to Causal Sets and Lattices

Another approach to non-regular geometry and quantum geometry are *causal sets* (see for example [1]), where the main structure is a transitive partial order relation $\leq$ with the interpretation of "lies in the future of." Given a causal fermion system, such a partial order relation can be defined by

$$x \leq y \qquad \text{if } x = y \text{ or } d(x,y) > 0$$

(where $d(x,y)$ is again the Lorentzian distance (13)). By construction, this relation is reflexive and transitive. However, it is not necessarily anti-symmetric in the sense that $x \leq y$ and $y \leq x$ implies $x = y$. Indeed, the above partial order might be trivial in the sense that it holds for all pairs $x, y \in M$. This is the case in particular if the causal fermion system is constructed starting from a Lorentzian manifold with closed timelike curves.

Having the partial order relation, one can follow the constructions in [2] and introduce an *orthogonality relation* by

$$x \perp y \qquad \text{if } x \not\leq y \text{ and } y \not\leq x$$

and setting

$$A^\perp = \{ x \in M \mid x \perp a \text{ for all } a \in A \} .$$

Likewise, the sets of subsets of M

$$\mathcal{L}(M, \perp) := \{ A \subset M \mid A^{\perp\perp} = A \}$$

forms a *lattice.*

5.3. A Riemannian Metric on $\mathcal{F}$

We finally introduce another inherent structure which has not been used so far but which might be helpful or of interest for geometers. To this end, let us again assume that our causal fermion system $(\mathcal{H}, \mathcal{F}, \rho)$ is regular (see Definition 3.1). Then we can also redefine $\mathcal{F}$ as the set of all self-adjoint operators on $\mathcal{H}$ which have exactly n positive and exactly n negative eigenvalues. This has the advantage that $\mathcal{F}$ has a smooth structure. In order to keep the setting reasonably simple, we restrict attention to the case that $\mathcal{H}$ is finite-dimensional, in which case $\mathcal{F}$ is a smooth manifold (in the infinite-dimensional setting, $\mathcal{F}$ would be a Banach manifold). Next, on $\mathcal{F}$ the Hilbert-Schmidt norm gives a distance function

$$d \,:\, \mathcal{F} \times \mathcal{F} \rightarrow \mathbb{R}_0^+ \,, \qquad d(x,y) = \|x-y\|_{\text{HS}} := \sqrt{\text{tr}\left((x-y)^2\right)}$$

(note that the existence of the trace is not an issue even in the infinite-dimensional setting because all operators in $\mathcal{F}$ have finite rank). The square of this distance function is smooth. Moreover, its first derivative vanishes on the diagonal, i.e. $D(d(x,.)^2)|_x = 0$. Therefore, taking its quadratic Taylor expansion about a point $x \in M$ gives a scalar product on $T_x\mathcal{F}$, i.e.

$$h_x \,:\, T_x\mathcal{F} \times T_x\mathcal{F} \rightarrow \mathbb{R} \,, \qquad h_x(u,v) = \text{tr}(uv) \,.$$

Clearly, this mapping depends smoothly on x and thus defines a *Riemannian metric* on $\mathcal{F}$.

IOP Conf. Series: Journal of Physics: Conf. Series **968** (2018) 012004 doi:10.1088/1742-6596/968/1/012004

Since every manifold can be endowed with a Riemannian metric, the only point of this construction is the fact that the above Riemannian metric is *canonical* in the sense that there is a distinguished Riemannian metric. This Riemannian metric could be useful for different purposes. Just to give an example, we here mention that it can be used for a *gauge-fixing procedure*: In computations (like the perturbation expansion in [6]), one often needs to expand functions on $\mathcal{F}$ in a Taylor expansion about a point $p \in \mathcal{F}$. Such an expansion is performed in a chart. The expansion coefficients of second and higher order depend in a complicated way on the choice of the chart. The freedom in choosing the charts includes the local gauge freedom in physics (for details see [6, Section 6.2]). Using the Riemannian metric h, the Taylor expansion can be performed in the distinguished chart given by the exponential map

$$\exp_x : \mathrm{U} \subset T_x\mathcal{F} \to \mathcal{F} .$$

In these charts, the gauge freedom is fixed completely.

Acknowledgments

I would like to thank the organizers of the workshop "Non-regular space-time geometry" held in Florence, June 2017, for the kind invitation and financial support. I am grateful to Valter Moretti and Clemens Sämann for helpful discussions.

References

[1] L. Bombelli, J. Lee, D. Meyer, and R.D. Sorkin, *Space-time as a causal set*, Phys. Rev. Lett. **59** (1987), no. 5, 521–524.
[2] W. Cegła, B. Jancewicz, and J. Florek, *Orthomodular lattice in Lorentzian globally hyperbolic space-time*, Rep. Math. Phys. **79** (2017), no. 2, 187–195.
[3] F. Finster, *A formulation of quantum field theory realizing a sea of interacting Dirac particles*, arXiv:0911.2102 [hep-th], Lett. Math. Phys. **97** (2011), no. 2, 165–183.
[4] ———, *Perturbative quantum field theory in the framework of the fermionic projector*, arXiv:1310.4121 [math-ph], J. Math. Phys. **55** (2014), no. 4, 042301.
[5] ———, *The Continuum Limit of Causal Fermion Systems*, arXiv:1605.04742 [math-ph], Fundamental Theories of Physics, vol. 186, Springer, 2016.
[6] ———, *Perturbation theory for critical points of causal variational principles*, arXiv:1703.05059 [math-ph] (2017).
[7] F. Finster and A. Grotz, *A Lorentzian quantum geometry*, arXiv:1107.2026 [math-ph], Adv. Theor. Math. Phys. **16** (2012), no. 4, 1197–1290.
[8] F. Finster, A. Grotz, and D. Schiefeneder, *Causal fermion systems: A quantum space-time emerging from an action principle*, arXiv:1102.2585 [math-ph], Quantum Field Theory and Gravity (F. Finster, O. Müller, M. Nardmann, J. Tolksdorf, and E. Zeidler, eds.), Birkhäuser Verlag, Basel, 2012, pp. 157–182.
[9] F. Finster and N. Kamran, *Spinors on singular spaces and the topology of causal fermion systems*, arXiv:1403.7885 [math-ph], to appear in Mem. Amer. Math. Soc. (2019).
[10] F. Finster and J. Kleiner, *Causal fermion systems as a candidate for a unified physical theory*, arXiv:1502.03587 [math-ph], J. Phys.: Conf. Ser. **626** (2015), 012020.
[11] ———, *A Hamiltonian formulation of causal variational principles*, arXiv:1612.07192 [math-ph], Calc. Var. Partial Differential Equations **56:73** (2017), no. 3, 33.
[12] F. Finster, J. Kleiner, and J.-H. Treude, *An Introduction to the Fermionic Projector and Causal Fermion Systems*, in preparation.
[13] F. Finster and O. Müller, *Lorentzian spectral geometry for globally hyperbolic surfaces*, arXiv:1411.3578 [math-ph], Adv. Theor. Math. Phys. **20** (2016), no. 4, 751–820.
[14] F. Finster and M. Reintjes, *A non-perturbative construction of the fermionic projector on globally hyperbolic manifolds I – Space-times of finite lifetime*, arXiv:1301.5420 [math-ph], Adv. Theor. Math. Phys. **19** (2015), no. 4, 761–803.
[15] M. Kunzinger and C. Sämann, *Lorentzian length spaces*, arXiv:1711.08990 [math.DG] (2017).

IOP Conf. Series: Journal of Physics: Conf. Series **968** (2018) 012005 doi:10.1088/1742-6596/968/1/012005

The Lorentzian distance formula in noncommutative geometry

Nicolas Franco

Namur Center for Complex Systems (naXys) & Department of Mathematics, University of Namur, rue de Bruxelles 61, 5000 Namur, Belgium

E-mail: nicolas.franco@unamur.be

Abstract. For almost twenty years, a search for a Lorentzian version of the well-known Connes' distance formula has been undertaken. Several authors have contributed to this search, providing important milestones, and the time has now come to put those elements together in order to get a valid and functional formula. This paper presents a historical review of the construction and the proof of a Lorentzian distance formula suitable for noncommutative geometry.

1. Introduction and formulation of the Lorentzian distance

Connes' noncommutative geometry [1, 2] provides at the same time a beautiful mathematical theory as well as new tools for physical models of unification theory. At a mathematical level, the topological correspondence between locally compact Hausdorff spaces and commutative C^*-algebras given by Gel'fand's theory is brought up to the level of Riemannian manifolds. The key elements are spectral triples $(\mathcal{A}, \mathcal{H}, D)$ from which, among others, information concerning the metric aspect can be recovered using the Riemannian distance formula :

$$d_R(p,q) \;=\; \sup_{f\in\mathcal{A}} \left\{ \, |f(q) - f(p)| \;:\; \|[D,f]\| \leq 1 \, \right\} \cdot \qquad (1)$$

Applications of noncommutative geometry in mathematical physics take part mainly in particle physics and quantum field theory. However the physical Lorentzian signature of spacetimes makes the use of the initial mathematical theory more problematic, especially concerning the formula (1). Two paths have been followed to solve this problem: the Wick rotation process which allows the use of all the well-defined tools of Riemannian noncommutative geometry (see e.g. [3, 4]), but therefore with a loss of causal information, or the adaptation of the theory to a Lorentzian signature, which is less straightforward and still ongoing, notably with the use of Krein spaces and Lorentzian spectral triples [5, 6, 7, 8]. In this last context, several authors have tried to generalize the formula (1) to a Lorentzian distance formula [9, 10, 11, 12, 13, 14, 15, 16]. Each of those authors has significantly contributed to a specific step of the construction of a final formula. In this paper, we will go through the summary of those different steps and we present two formulations of a now completely proved Lorentzian distance formula.

The first formulation is at the level of traditional Lorentzian geometry, where the usual Lorentzian distance $d(p, q)$ between two points, representing the maximal length of the piecewise C^1 future-directed causal curves from p to q [17], is rewritten in a completely path-independent

IOP Conf. Series: Journal of Physics: Conf. Series **968** (2018) 012005 doi:10.1088/1742-6596/968/1/012005

way, using the information coming from a specific set of test functions. Key elements of the proof of the following formula will be presented in Section 3.

Theorem 1 (Path-independent formulation).
If (M, g) is a time-oriented Lorentzian manifold (spacetime) which is either:

- *globally hyperbolic,*
- *stably causal such that the usual Lorentzian distance d is continuous and finite,*

then for all $p, q \in M$:

$$d(p,q) \;=\; \inf \left\{ \; [f(q) - f(p)]^+ \; : \; f \in \mathscr{S} \; \right\}, \tag{2}$$

where $[\alpha]^+ = \max\{0, \alpha\}$ and $\mathscr{S}$ is the set of smooth real-valued "steep" functions, i.e. the set of $f \in C^1(M, \mathbb{R})$ such that $g(\nabla f, \nabla f) = g^{-1}(df, df) \leq -1$ and ∇f is past-directed (f is a future-directed temporal function).

Stable causality is the weakest assumption under which the RHS of (2) makes sense, otherwise the set of steep functions $\mathscr{S}$ is empty [18]. The condition of continuity of the Lorentzian distance d is necessary since the RHS is upper semi-continuous while the LHS is lower semi-continuous [15]. Under such an assumption, (M, g) is in fact a causally continuous spacetime in the sense of [19]. The condition of global hyperbolicity is a particular case where the Lorentzian distance d is automatically continuous and finite [17].

The second formulation is an algebraic formulation, where every element from traditional Lorentzian geometry has been replaced by a corresponding element coming from the theory of spectral triples. This formulation opens the possibility of a generalization to noncommutative spacetimes. The proof will be presented in Section 4 while the possible technical difficulties for an application on noncommutative spacetimes will be presented in Section 5.

Theorem 2 (Spectral triple formulation).
If (M, g) is a n-dimensional spin Lorentzian manifold which is either globally hyperbolic or stably causal such that the Lorentzian distance d is continuous and finite, and if we define:

- *The algebra $\mathcal{A} = C^1(M, \mathbb{R})$ with pointwise multiplication,*
- *The Hilbert space $\mathcal{H} = L^2(M, S)$ of square integrable sections of the spinor bundle S over M (using a positive definite inner product on the spinor bundle),*
- *The Dirac operator $D = -i(\hat{c} \circ \nabla^S) = -ie^\mu_a \gamma^a \nabla^S_\mu$ associated with the spin connection ∇^S,*
- *The fundamental symmetry $\mathcal{J} = i\gamma^0$ where γ^0 is the first flat gamma matrix*[1]*,*
- *If n is even, the chirality operator $\chi = \pm i^{\frac{n}{2}+1}\gamma^0 \cdots \gamma^{n-1}$,*

then for all $p, q \in M$, if n is even:

$$d(p,q) \;=\; \inf_{f \in \mathcal{A}} \left\{ \; [f(q) - f(p)]^+ \; : \; \forall \phi \in \mathcal{H}, \langle \phi, \mathcal{J}([D, f] + i\chi)\phi \rangle \leq 0 \; \right\}, \tag{3}$$

and if n is odd:

$$d(p,q) \;=\; \inf_{f \in \mathcal{A}} \left\{ \; [f(q) - f(p)]^+ \; : \; \forall \phi \in \mathcal{H}, \langle \phi, \mathcal{J}([D, f] \pm 1)\phi \rangle \leq 0 \; \right\}, \tag{4}$$

where $[\alpha]^+ = \max\{0, \alpha\}$ and $\langle \cdot, \cdot \rangle$ is the positive definite inner product on $\mathcal{H}$.

[1] Conventions used in the paper are $(-, +, +, +, \cdots)$ for the signature of the metric and $\{\gamma^a, \gamma^b\} = 2\eta^{ab}$ for the flat gamma matrices, with γ^0 anti-Hermitian and γ^a Hermitian for $a > 0$.

2. Historical construction of the Riemannian and Lorentzian distance formulas

The first apparition of Connes' distance formula (1) comes from 1989 in [20]. Common presentations and proofs of the formula are given by [21], [1, Chapter 6] and [22, Chapter 9.3]. The formula has been studied and applied by many authors on several kinds of spaces, as among others [23, 24, 25, 26, 27, 28, 29].

The way to prove this formula is quite direct. If we consider a connected compact Riemannian manifold (M, g) and two points p and q on it, we can choose an arbitrary piecewise C^1 curve $\gamma : [0,1] \to M$ with $\gamma(0) = p$ and $\gamma(1) = q$. Then, for each function $f \in C^\infty(M)$, we have by using the second fundamental theorem of calculus:

$$\begin{aligned} f(q) - f(p) &= f(\gamma(1)) - f(\gamma(0)) = \int_0^1 \frac{\mathrm{d}}{\mathrm{d}t} f(\gamma(t))\, dt \\ &= \int_0^1 df(\dot\gamma(t))\, dt = \int_0^1 g(\nabla f, \dot\gamma(t))\, dt. \end{aligned}$$

Using the Cauchy–Schwarz inequality, we get:

$$\begin{aligned} |f(q) - f(p)| &\leq \int_0^1 |g(\nabla f, \dot\gamma(t))|\, dt \leq \int_0^1 |\nabla f|\, |\dot\gamma(t)|\, dt \\ &\leq \|\nabla f\|_\infty \int_0^1 |\dot\gamma(t)|\, dt = \|\nabla f\|_\infty\, l(\gamma), \end{aligned} \tag{5}$$

where $l(\gamma)$ denotes the length of the curve. So we obtain the following inequality:

$$d_R(p,q) \geq \sup \left\{ |f(q) - f(p)| \ :\ f \in \mathcal{A},\ \|\nabla f\|_\infty \leq 1 \right\}. \tag{6}$$

The condition $\|\nabla f\|_\infty \leq 1$ can be replaced by a weaker condition $\operatorname{ess\,sup} \|\nabla f\| \leq 1$ which allows us to work with the set $\mathcal{A} \subset C(M)$ of Lipschitz continuous functions on M [30]. Within this larger set, the equality is easily given by the usual distance as a function of its second argument $f(\cdot) = d_R(p,\cdot)$. Indeed, d_R is Lipschitz continuous with $\|\nabla d_R\| = 1$ except on a set of measure zero (the point p and the cut locus), and we get the path-independent formula:

$$d_R(p,q) = \sup \left\{ |f(q) - f(p)| \ :\ f \in \mathcal{A},\ \operatorname{ess\,sup} \|\nabla f\| \leq 1 \right\}. \tag{7}$$

The last step is the translation of this formula into an algebraic formalism. If M is a spin manifold and D the Dirac operator, we have [22, Chapter 9.3]:

$$\operatorname{ess\,sup} \|\nabla f\| \leq 1 \iff \|[D, f]\| \leq 1 \tag{8}$$

which gives the formula (1).

The construction of the Riemannian distance formula can be clearly divided in three important steps: the setting of a path-independent inequality (6), the construction of the equality case (7) and the operatorial (spectral triple) formulation (8). The search for a Lorentzian equivalent formula went through the same three steps and we summarize here its historical evolution:

- 1998-2000, G. N. Parfionov and R. R. Zapatrin [9]: First mention of the duality (inversion supremum-infimum and the inequality signs) in the formula (6) in a Lorentzian context.
- 2002-2003, V. Moretti [10]: Generalization of the formula (7) for globally hyperbolic spacetimes using a local condition on the gradient ∇f (in a more recent terminology: using functions that are "steep" almost-everywhere but only inside some specific compact sets) and an attempt at algebraization using the Laplace-Beltrami-d'Alembert operator and a net of Hilbert spaces.

IOP Conf. Series: Journal of Physics: Conf. Series **968** (2018) 012005 doi:10.1088/1742-6596/968/1/012005

- 2010, N. Franco [12]: Generalization of the formula (7) for globally hyperbolic spacetimes using a global condition on the gradient ∇f (using functions that are steep almost-everywhere on the whole spacetime). The global behavior of the test functions is chosen in order to facilitate a future algebraization. The proof of the equality case is done using non-Lipschitz continuous causal functions.
- 2012-2013, N. Franco and M. Eckstein [13]: Algebraic formulation of the global condition on the gradient (steep) for C^1 functions, so a Lorentzian generalization of (8). However, this algebraic formulation is not valid for non-Lipschitz continuous functions as needed for the general proof in [12], so the proof of the distance formula is limited to spacetimes where the usual distance function can be suitably approximated by C^1 steep functions. A particular proof for the Minkowskian case is given.
- 2014-2016, A. Rennie and B. E. Whale [14]: Extension of the formula obtained in [12] for non-globally hyperbolic spacetimes. The correspondence is extended to spacetimes where the usual Lorentzian distance is finite, while conjecturing that the condition of stable causality should be necessary if the distance is also continuous. The steep condition is also proved to be necessary for a formulation in term of test functions.
- 2017, E. Minguzzi [15]: As a consequence of the study of causality under less-regular differentiability with the smoothing of non-Lipschitz continuous steep functions, a C^1 proof of the formula given in [12] is obtained, with the necessary and sufficient condition that the spacetime must be stably causal and the usual Lorentzian distance function finite and continuous. This gives a complete smooth validation of the results presented in [12, 13, 14]. An additional C^1 proof of the formula, simplest but under the assumption of a globally hyperbolic spacetimes, is also presented in [16].

3. The path-independent formulation of the Lorentzian distance

In this Section, we summarize the main arguments of the proof of Theorem 1. The key elements of the Lorentzian distance formula are:

- The real-valued (continuous or not) causal functions, which are the functions which do not decrease along every future-directed causal curve.
- The steep functions, which are C^1 causal functions which increase sufficiently rapidly along every future-directed causal curve., i.e. with $g(\nabla f, \nabla f) \leq -1$ and past-directed gradient.

Unlike the Riemannian distance formula (1), we have to consider real-valued functions instead of complex ones in order to reach a non-symmetric formula.

At first, we need an inequality which is a Lorentzian generalization of (6). The same two theorems are used in their existing Lorentzian versions:

- The second fundamental theorem of calculus, valid for absolutely continuous, hence C^1, functions.
- The reverse Cauchy-Schwarz inequality [17]:

$$\text{If } v \text{ and } w \text{ are timelike vectors, then } |g(v,w)| \geq \sqrt{-g(v,v)}\sqrt{-g(w,w)}.$$

Now we consider a time-oriented Lorentzian manifold (M, g) and two points p and q on it such that $p \lll q$. We can choose a piecewise C^1 future-directed timelike curve $\gamma : [0,1] \to M$ with $\gamma(0) = p$ and $\gamma(1) = q$. Then, for each function $f \in C^1(M, \mathbb{R})$,

$$f(q) - f(p) = \int_0^1 \frac{\mathrm{d}}{\mathrm{d}t} f(\gamma(t))\, dt = \int_0^1 g(\nabla f, \dot{\gamma}(t))\, dt. \tag{9}$$

Since $\dot{\gamma}(t)$ is almost everywhere a future-directed timelike vector, if we assume that ∇f is everywhere timelike with constant past-directed orientation, then the sign of $g(\nabla f, \dot{\gamma}(t))$ is constant and positive, so we get:

$$\begin{aligned} f(q) - f(p) &= \int_0^1 g(\nabla f, \dot{\gamma}(t))\, dt = \int_0^1 |g(\nabla f, \dot{\gamma}(t))|\, dt \\ &\geq \int_0^1 \sqrt{-g(\nabla f, \nabla f)} \sqrt{-g(\dot{\gamma}(t), \dot{\gamma}(t))}\, dt \\ &\geq \inf \left\{ \sqrt{-g(\nabla f, \nabla f)} \right\} l(\gamma), \end{aligned} \tag{10}$$

which is the Lorentzian counterpart of (5). This result can be extended by continuity to future-directed causal curves and $p \preceq q$. Then taking the supremum over all future-directed causal curves from p to q we get the following path-independent inequality:

$$d(p,q) \leq \inf_{f \in C^1(M,\mathbb{R})} \left\{ [f(q) - f(p)]^+ \ : \ {\scriptstyle g(\nabla f, \nabla f) \leq -1, \atop \nabla f \text{ past-directed}} \right\}. \tag{11}$$

Note that this result can be extended to non-absolutely continuous functions (hence non-Lipschitz continuous) as long as we impose that the functions f remain causal. Indeed, from their monotony, those functions are a.e. differentiable on any future-directed causal curve $\gamma : [0,1] \to M$ and instead of (9) we have the inequality $f(\gamma(1)) - f(\gamma(0)) \geq \int_0^1 \frac{\mathrm{d}}{\mathrm{d}t} f(\gamma(t))\, dt$ which leads to the same formulation (10) and to the following formula:

$$d(p,q) \leq \inf_{\substack{\text{causal} \\ \text{functions } f}} \left\{ [f(q) - f(p)]^+ \ : \ {\scriptstyle \operatorname{ess\,sup} g(\nabla f, \nabla f) \leq -1, \atop \nabla f \text{ past-directed}} \right\}. \tag{12}$$

Three proofs have been given concerning the equality between the usual Lorentzian distance and the formulas (11) or (12). Due to the length of those technical proofs, we only present the main ideas here.

Under the condition of global hyperbolicity, a proof of the equality for the non-smooth formulation (12) is presented in [12], with the construction of a sequence of a.e. steep continuous functions converging to the equality and constructed as locally finite sums of distance functions computed from points located near a suitable Cauchy surface. This particular proof can now be transformed in a smooth version using the new results from [15], which is explicitly done in [16], but it is still limited to globally hyperbolic spacetimes due to the use of a Cauchy surface as main element. The second proof presented in [14] gets rid of the Cauchy surface and instead uses the construction of a specific achronal surface S such that $M = I^+(S) \cup S \cup I^-(S)$ and considers the distance $f(\cdot) = d(S, \cdot)$ to this surface as the equality function. This extends the proof of the non-smooth formulation (12) to spacetimes with finite Lorentzian distance (where the distance is also allowed to be non-continuous).

The most recent and general proof can be found in [15] and concerns the smooth formulation (11). It implies the precedent proofs (at least when the distance is continuous). The necessary and sufficient conditions are the stable causality of the spacetime and finiteness and continuity of the Lorentzian distance (which is automatic if the spacetime is globally hyperbolic). The approach is here completely different and uses the idea that metric properties of spacetimes can be computed using a causal theory on a space with one extra dimension $\tilde{M} = M \times \mathbb{R}$. The usual Lorentzian distance function can then be traded for a Lorentz-Finsler function defined on causal tangent vectors of the product space. The final proofs of Theorem 1 is then given by [15, Theorem 4.11] in the globally hyperbolic case and by [15, Theorem 4.15] in the more general case of stably causal spacetimes, requiring the additional condition of finiteness and continuity of the Lorentzian distance.

IOP Conf. Series: Journal of Physics: Conf. Series **968** (2018) 012005 doi:10.1088/1742-6596/968/1/012005

4. The algebraic formulation of the Lorentzian distance

In this Section, we present the proof of Theorem 2 and especially the origin of the algebraic constraint inside (3) and (4). Once more, we will see that the metric information emerges naturally from the causal information on a space with one extra dimension. So at first we need to review the causal theory for spectral triples.

We consider a stably causal Lorentzian n-dimensional manifold (M, g) with a spin structure S with its associated space $L^2(M, S)$ of square integrable sections of the spinor bundle over M. This space is naturally endowed with a indefinite inner product $(\cdot, \cdot)$ coming from the spin structure and possesses all the properties of a Krein space [31], but can be turned into a Hilbert space $\mathcal{H}$ using an alternative positive-definite inner product $\langle \cdot, \cdot \rangle = (\cdot, \mathcal{J} \cdot)$ and $(\cdot, \cdot) = \langle \cdot, \mathcal{J} \cdot \rangle$. The operator $\mathcal{J}$ is called a fundamental symmetry and can be constructed from the Clifford action of a timelike vector field [32]. Since stable causality implies the existence of a smooth temporal function $\mathcal{T}$ [18], a (Hermitian) fundamental symmetry is easily given by $\mathcal{J} = ic(d\mathcal{T}) = i\gamma^0$, where γ^0 is the first flat gamma matrix respecting $(\gamma^0)^2 = -1$ (up to a smooth conformal transformation which leaves the causality invariant). If we consider the Lorentzian Dirac operator $D = -ie_a^\mu \gamma^a \nabla_\mu^S$, where e_a^μ stand for the vierbeins[2], then this operator is anti-symmetric for the Krein product, which is equivalent to say that $\mathcal{J}D$ is an anti-symmetric operator for the Hilbert space $\mathcal{H}$. Under the additional assumption of completeness of the manifold M under spacelike reflexion, $\mathcal{J}D$ is a skew-Hermitian operator on $\mathcal{H}$ [5].

Then we have the following theorem coming from [13] :

Theorem 3. *A function $f \in C^1(M, \mathbb{R})$ is causal if and only if*

$$\forall \phi \in \mathcal{H}, \langle \phi, \mathcal{J}[D, f]\phi \rangle \leq 0$$

where $\langle \cdot, \cdot \rangle$ is the positive definite inner product on $\mathcal{H}$.

Two elements are important for the proof of this Theorem. At first, the absolute continuity of the function f. If the function is absolutely continuous, then the causal property can be fully characterized by the following conditions on its gradient:

$$g(\nabla f, \nabla f) = g^{\mu\nu} f_{,\mu} f_{,\nu} \leq 0, \qquad g(\nabla f, \nabla \mathcal{T}) = g^{\mu 0} f_{,\mu} = -f_{,0} \leq 0, \tag{13}$$

where $df = f_{,\mu} dx^\mu$ and $x^0 = \mathcal{T}$ is orthonormal to the others chosen local coordinates. The second element is the C^1 behavior. Indeed, from the continuity of the derivative, if (13) is false at some point of M, then it must be false on some neighborhood and this information can be caught by a suitable specific spinor. From this observation, we can see that the smooth formulation of the Lorentzian distance equation (2) is important for the algebraic generalization.

From the well-known property $[D, f] = -i\, c(df)$ [22], the proof of Theorem 3 relies on the fact that the matrix:

$$\mathcal{J}[D, f] = i\gamma^0 \,(-i)\, (\gamma^a e_a^\mu f_{,\mu}) = \gamma^0 \gamma^a e_a^\mu f_{,\mu}$$

is pointwise negative semi-definite if and only if (13) is respected. A first and technical proof of this equivalence is given in [13] using the technique of the characteristic polynomial.[3] We present here another shorter consideration suggested in [33]. We have

$$\mathcal{J}[D, f] = -f_{,0} + \gamma^0 \gamma^i e_i^\mu f_{,\mu} = -f_{,0} + b$$

[2] We consider here a "pseudo-orthonormal" frame coming from the timelike vector field $\partial_0 = \partial_{\mathcal{T}}$ with $e_0^0 = 1$ and $e_0^i = e_i^0 = 0$ for $i = 1, \ldots, n-1$.

[3] The initial proof was done under the hypothesis of global hyperbolicity but can be extended to simply causal spacetimes by considering the pseudo-orthonormal frame.

IOP Conf. Series: Journal of Physics: Conf. Series **968** (2018) 012005 doi:10.1088/1742-6596/968/1/012005

for $i = 1, \ldots, n-1$ where the second term b is Hermitian and respects $b^2 = \left\|\gamma^i\gamma^j f_{,i}f_{,j}\right\|^2$. So the spectrum of b must be $\{\pm \left\|g^{ij} f_{,i}f_{,j}\right\|\}$. Since the reduced metric g^{ij} is positive definite we get that $\mathcal{J}[D,f] = -f_{,0} + b$ is negative semi-definite if and only if $-f_{,0} \leq 0$ and $g^{ij}f_{,i}f_{,j} \leq f_{,0}^2$, hence (13).

Theorem 3 is very important in noncommutative geometry since its provides a way to completely characterize causality at an algebraic level, since the complete set of causal functions is sufficient to characterize the causal relations for stably causal spacetimes [15, 34]. Consequences of Theorem 3 on noncommutative spacetimes (almost commutative spacetimes, Moyal spacetime) have already been handled with success [29, 35, 36, 37, 38, 39].

From Theorem 3, we can now easily get the characterization of the steep functions used in (2). Once more, we consider a product space $\tilde{M} = M \times \mathbb{R}$ on which we extend the Lorentzian metric g to $\tilde{g}$ by adding $\tilde{g}^{nn} = 1$ and $\tilde{g}^{\mu n} = \tilde{g}^{n\mu} = 0$ for $\mu = 0, \ldots, n-1$. We will use the extended indices $\tilde{a}, \tilde{\mu} = 0, \ldots, n$. We also have to extend the spin structure to the new dimension, giving an extended Dirac operator $\tilde{D}$. When n is even, this can be done very easily by considering the chirality operator as an additional gamma matrix $\gamma^n = \pm\chi = \pm i^{\frac{n}{2}+1}\gamma^0 \cdots \gamma^{n-1}$ (with $e^\mu_n = e^n_a = 0$ and $e^n_n = 1$). Now we can consider all functions of the form $\tilde{f} = f - x_n \in C^1(\tilde{M}, \mathbb{R})$ where $f \in C^1(M, \mathbb{R})$, which trivially gives $\tilde{f}_{,\mu} = f_{,\mu}$ and $\tilde{f}_{,n} = -1$.

Then applying Theorem 3 to $\tilde{f}$ gives:

$$\begin{aligned}
\forall \phi \in \mathcal{H}, \qquad & \left\langle \phi, \mathcal{J}[\tilde{D}, \tilde{f}]\phi \right\rangle \\
&= \left\langle \phi, \mathcal{J}\left(-i\gamma^{\tilde{a}} e^{\tilde{\mu}}_{\tilde{a}} \tilde{f}_{,\tilde{\mu}}\right)\phi \right\rangle \\
&= \left\langle \phi, \mathcal{J}\left(-i\gamma^{a} e^{\mu}_{a} f_{,\mu} - i\gamma^n \tilde{f}_{,n}\right)\phi \right\rangle \\
&= \left\langle \phi, \mathcal{J}\left([D,f] \pm i\chi\right)\phi \right\rangle \;\leq\; 0
\end{aligned}$$

which is equivalent to the fact that $\tilde{f}$ is causal on $(\tilde{M}, \tilde{g})$, i.e. $-\tilde{f}_{,0} = -f_{,0} \leq 0$ and:

$$\tilde{g}(\tilde{\nabla}\tilde{f}, \tilde{\nabla}\tilde{f}) = \tilde{g}^{\tilde{\mu}\tilde{\nu}} \tilde{f}_{,\tilde{\mu}} \tilde{f}_{,\tilde{\nu}} = g^{\mu\nu} f_{,\mu} f_{,\nu} + \tilde{g}^{nn} \tilde{f}_{,n} \tilde{f}_{,n} = g(\nabla f, \nabla f) + 1 \leq 0$$

which is the exact characterization of a steep function. The choice of $+i\chi$ or $-i\chi$ is completely arbitrary and has no influence on the formula, so from (2) we can write:

$$d(p,q) \;=\; \inf_{f \in C^1(M,\mathbb{R})} \left\{ \;[f(q) - f(p)]^+ \;:\; \forall \phi \in \mathcal{H}, \langle \phi, \mathcal{J}([D,f] + i\chi)\phi \rangle \leq 0 \;\right\},$$

which is valid for manifolds with even dimension n respecting the conditions of Theorem 1. So the proof of the formula (3) is complete.

This process can also be applied in order to get a valid formula for odd-dimensional manifolds, but this requires a doubling of the Hilbert space $\tilde{\mathcal{H}} = \mathcal{H} \otimes \mathbb{C}^2$ and new gamma matrices $\tilde{\gamma}^\mu = \gamma^\mu \otimes \sigma^1$ for $\mu = 0, \ldots, n-1$ and $\tilde{\gamma}^n = 1 \otimes \sigma^2$ where σ^i are the Pauli matrices. The fundamental symmetry becomes $\tilde{\mathcal{J}} = \mathcal{J} \otimes \sigma^1$. The negative semi-definite operator becomes:

$$\begin{aligned}
\tilde{\mathcal{J}}[\tilde{D}, \tilde{f}] &= (\mathcal{J} \otimes \sigma^1)\left(-ie^\mu_a f_{,\mu}(\gamma^a \otimes \sigma^1) + i(1 \otimes \sigma^2)\right) \\
&= \mathcal{J}[D,f] \otimes 1 + \mathcal{J} \otimes \sigma^3.
\end{aligned}$$

However, $\sigma^3 = \left(\begin{smallmatrix} 1 & 0 \\ 0 & -1 \end{smallmatrix}\right)$ is diagonal, which means that the constraint:

$$\forall \tilde{\phi} = (\phi_+, \phi_-) \in \tilde{\mathcal{H}}, \; \left\langle \tilde{\phi}, \tilde{\mathcal{J}}[\tilde{D}, \tilde{f}]\tilde{\phi} \right\rangle \leq 0$$

splits into two inequalities:

$$\forall \phi_\pm \in \mathcal{H}, \; \langle \phi_\pm, \mathcal{J}([D,f] \pm 1)\phi_\pm \rangle \leq 0.$$

This gives rise to formula (4) and completes the proof of Theorem 2.

IOP Conf. Series: Journal of Physics: Conf. Series **968** (2018) 012005 doi:10.1088/1742-6596/968/1/012005

5. Application on noncommutative spacetimes

Theorem 2 opens the possibility to compute metrical information on noncommutative spacetimes. The formalism is the one of Lorentzian spectral triples $(\mathcal{H}, \mathcal{A}, D)$ with a given fundamental symmetry $\mathcal{J}$ with $\mathcal{J}^2 = 1$, $\mathcal{J}^* = \mathcal{J}$, $[\mathcal{J}, a] = 0$, $\forall a \in \mathcal{A}$. The conditions on the operator D are $\forall a \in \mathcal{A}$, $[D, a]$ is bounded, $a(1 + \frac{(DD^*+D^*D)}{2})^{-\frac{1}{2}}$ is compact and $D^* = -\mathcal{J}D\mathcal{J}$ (Krein skew-selfadjoint). For an even Lorentzian spectral triple, the $\mathbb{Z}_2$-grading χ must respect $\chi^* = \chi$, $\chi^2 = 1$, $[\chi, a] = 0$, $\chi\mathcal{J} = -\mathcal{J}\chi$ and $\chi D = -D\chi$. The compactness condition on $a(1+\frac{(DD^*+D^*D)}{2})^{-\frac{1}{2}}$ is not necessary for the metrical information but it is a natural generalization of the usual compact resolvent condition coming from the Riemannian case.

Then a Lorentzian distance, respecting the usual properties of non-negativity, antisymmetry and inverse triangle inequality, can be defined between two states φ, ψ on $\mathcal{A}$ by:

$$d(\varphi, \psi) = \inf_{a \in \mathcal{A}} \left\{ [\psi(a) - \varphi(a)]^+ \ : \ \forall \phi \in \mathcal{H}, \langle \phi, \mathcal{J}([D, a] + i\chi)\phi \rangle \leq 0 \right\}, \tag{14}$$

where $i\chi$ should be replaced by ± 1 (both signs) if the Lorentzian spectral triple is odd.

There exist two technical difficulties in applying the formula (14) which we are going to discuss. However, we will show that in all currently existing examples, those difficulties can be bypassed.

The first difficulty concerns the fundamental symmetry $\mathcal{J}$. With the minimal set of axioms presented here concerning $\mathcal{J}$, there is no guarantee that the signature is exactly Lorentzian (it can correspond to a pseudo-Riemannian manifold in the commutative case) which means that the Lorentzian distance formula could give no result. The exact set of axioms in order to guarantee a Lorentzian signature is still an active subject of research, with some existing but not identical working proposals [6, 7, 8, 36]. Since we want to keep this paper as general as possible, we will not focus on one particular condition. The existing typical examples of noncommutative spacetimes, all of which could fit into the Lorentzian spectral triple formalism, are almost-commutative manifolds (Kaluza-Klein products of a usual Lorentzian manifold and a discrete noncommutative internal space [35, 37]) and deformations of flat spacetimes (Moyal spacetime, κ-Minkowski, Lorentzian cylinder, ... [29, 40]). For almost commutative manifolds, the suitable fundamental symmetry is $\mathcal{J} = \mathcal{J}_M \otimes 1$ where $\mathcal{J}_M$ is a fundamental symmetry for the based spacetime. For deformations of flat spacetimes, the canonical choice is $\mathcal{J} = i\gamma^0$. So, to our knowledge, there is no currently existing toy model (noncommutative noncompact complete Lorentzian spacetime) for which this problem must be taken into consideration, and the question is reserved for abstract considerations.

The second difficulty concerns the algebra $\mathcal{A}$ and the space of states on it. In traditional noncommutative geometry, $\mathcal{A}$ is a pre-C*-algebra, since it corresponds in the commutative case to continuous functions vanishing at infinity. For causal considerations (causal functions), this algebra is too small and one must consider an additional specific unitization of $\mathcal{A}$ corresponding to bounded functions [13]. However, steep functions present in the Lorentzian distance formula are clearly unbounded and cannot fit into the usual pre-C*-algebra formalism. One must find a way to extend the initial pre-C*-algebra corresponding to the states to unbounded elements and be sure that (some of) the states are still well-defined and uniquely extended. One particular way to realize such an extension is presented in [6]. Again, for all currently existing examples, the problem can be at least partially bypassed. For almost commutative manifolds, all pure states are product states between well-defined states on the based spacetime (evaluation maps) and vector states on the discrete algebra. For deformation spaces, the elements of the initial pre-C*-algebra of bounded continuous functions are compact operators so all states correspond to vector states which can be easily and uniquely extended to unbounded functions as long as their evaluation remain finite, as used in [29]. Once more, this problem is an abstract one and does not prevent the application of the Lorentzian distance formula to particular models of noncommutative spacetimes.

Acknowledgments
The author would like to thank Ettore Minguzzi for organizing the meeting "Non-Regular Spacetime Geometry" between the two communities of Lorentzian geometry and noncommutative geometry, from which the evolution of the proof of the Lorentzian distance formula was possible.

References

[1] Connes A 1994 *Noncommutative Geometry* (San Diego: Academic Press) ISBN 978-0-1218-5860-5
[2] Connes A and Marcolli M 2008 *Noncommutative Geometry, Quantum Fields and Motives* (*Colloquium Publications* vol 55) (American Mathematical Society) ISBN 978-0-8218-4210-2
[3] Schücker T 2005 Forces from connes' geometry *Topology and Geometry in Physics* ed Bick E and Steffen F D (Springer Berlin Heidelberg) pp 285–350 ISBN 978-3-540-31532-2 (*Preprint* hep-th/0111236)
[4] D'Andrea F, Kurkov M A and Lizzi F 2016 *Phys. Rev. D* **94**(2) 025030 (*Preprint* 1605.03231)
[5] Strohmaier A 2006 *Journal of Geometry and Physics* **56** 175–195 (*Preprint* math-ph/0110001)
[6] Franco N 2014 *Reviews in Mathematical Physics* **26** 1430007 (*Preprint* 1210.6575)
[7] van den Dungen K, Paschke M and Rennie A 2013 *Journal of Geometry and Physics* **73** 37–55 (*Preprint* 1207.2112)
[8] Besnard F and Bizi N 2017 *Journal of Geometry and Physics* (*Preprint* 1611.07830)
[9] Parfionov G N and Zapatrin R R 2000 *Journal of Mathematical Physics* **41** 7122–7128
[10] Moretti V 2003 *Reviews in Mathematical Physics* **15** 1171–1217 (*Preprint* gr-qc/0203095)
[11] Franco N 2010 *AIP Conference Proceedings* **1241** 588–594 (*Preprint* 1003.5407)
[12] Franco N 2010 *Symmetry, Integrability and Geometry: Methods and Applications* **6** 064 special Issue "Noncommutative Spaces and Fields" (*Preprint* 1003.5651)
[13] Franco N and Eckstein M 2013 *Classical and Quantum Gravity* **30** 135007 (*Preprint* 1212.5171)
[14] Rennie A and Whale B E 2016 *Journal of Geometry and Physics* **106** 108 – 121 (*Preprint* 1412.5652)
[15] Minguzzi E 2017 Causality theory for closed cone structures with applications preprint arXiv:1709.06494
[16] Minguzzi E 2017 The representation of spacetime through steep time functions preprint arXiv:1711.00076
[17] Beem J, Ehrlich P and Easley K 1996 *Global Lorentzian Geometry* (*Monographs and Textbooks in Pure and Applied Mathematics* vol 202) (CRC Press) ISBN 978-0-8247-9324-1
[18] Minguzzi E and Sánchez M 2008 The causal hierarchy of spacetimes *Recent Developments in Pseudo-Riemannian Geometry, ESI Lectures in Mathematics and Physics* ed Alekseevsky D V and Baum H (European Mathematical Society Publishing House) pp 299–358 ISBN 978-3-03719-051-7 (*Preprint* gr-qc/0609119)
[19] Hawking S W and Ellis G F R 1975 *The Large Scale Structure of Space-Time (Cambridge Monographs on Mathematical Physics)* (Cambridge University Press) ISBN 0521099064
[20] Connes A 1989 *Ergodic Theory and Dynamical Systems* **9** 207–220
[21] Connes A and Lott J 1992 *The Metric Aspect of Noncommutative Geometry* (Boston, MA: Springer US) pp 53–93
[22] Gracia-Bondía J M, Várilly J C and Figueroa H 2001 *Elements of Noncommutative Geometry* (Springer) ISBN 978-1-4612-0005-5
[23] Bimonte G, Lizzi F and Sparano G 1994 *Physics Letters B* **341** 139 – 146 (*Preprint* hep-lat/9404007)
[24] Rieffel M A 1999 *Documenta Mathematica* **4** 559–600 (*Preprint* math/9906151)
[25] Iochum B, Krajewski T and Martinetti P 2001 *Journal of Geometry and Physics* **37** 100–125 (*Preprint* hep-th/9912217)
[26] Cagnache E and Wallet J C 2010 *SIGMA* **6** 17 (*Preprint* 0912.4185)
[27] Cagnache E, D'Andrea F, Martinetti P and Wallet J C 2011 *Journal of Geometry and Physics* **61** 1881–1897 (*Preprint* 0912.0906)
[28] Martinetti P and Tomassini L 2013 *Communications in Mathematical Physics* **323** 107–141 (*Preprint* 1110.6164)
[29] Franco N and Wallet J C 2016 Metrics and causality on moyal planes *Noncommutative Geometry and Optimal Transport* (*Contemporary Mathematics* vol 676) ed Martinetti P and Wallet J C (American Mathematical Society) pp 147–173 ISBN 978-1-4704-2297-4 (*Preprint* 1507.06559)
[30] Weaver N 1996 *Journal of Functional Analysis* **139** 261–300
[31] Bognár J 1974 *Indefinite Inner Product Spaces* (Springer) ISBN 978-3-642-65569-2
[32] Müller O and Sánchez M 2014 *Jahresbericht der Deutschen Mathematiker-Vereinigung* **115** 153–183 (*Preprint* math/0604265)
[33] Besnard F 2015 *Journal of Physics: Conference Series* **634** 012009 (*Preprint* 1508.01917)
[34] Besnard F 2009 *Journal of Geometry and Physics* **59** 861–875 (*Preprint* 0804.3551)

[35] Franco N and Eckstein M 2014 *Symmetry, Integrability and Geometry: Methods and Applications* **10** 010 special Issue on Noncommutative Geometry and Quantum Groups in honor of Marc A. Rieffel. (*Preprint* 1310.8225)
[36] Franco N and Eckstein M 2014 Noncommutative geometry, Lorentzian structures and causality *Mathematical Structures of the Universe* ed Eckstein M, Heller M and Szybka S (Copernicus Center Press) pp 315–340 ISBN 978-8-3788-6107-2 (*Preprint* 1409.1480)
[37] Franco N and Eckstein M 2015 *Journal of Geometry and Physics* **96** 42–58 (*Preprint* 1502.04683)
[38] Eckstein M and Franco N 2015 Causal structure for noncommutative geometry *Frontiers of Fundamental Physics 14* 138 URL https://pos.sissa.it/224/138/pdf
[39] Eckstein M, Franco N and Miller T 2017 *Phys. Rev. D* **95**(6) 061701R (*Preprint* 1610.10083)
[40] van Suijlekom W D 2004 *Journal of Mathematical Physics* **45** 537–556 (*Preprint* math-ph/0310009)

IOP Conf. Series: Journal of Physics: Conf. Series **968** (2018) 012006 doi:10.1088/1742-6596/968/1/012006

Explicit triangular decoupling of the separated vector wave equation on Schwarzschild into scalar Regge-Wheeler equations

Igor Khavkine

Università di Milano, Via Cesare Saldini, 50, I-20133 Milano, Italy
Istituto Nazionale di Fisica Nucleare – Sezione di Milano, Via Celoria, 16, I-20133 Milan, Italy

E-mail: igor.khavkine@unimi.it

Abstract. We consider the vector wave equation on the Schwarzschild spacetime, which can be considered as coming from the harmonic (or Lorenz) gauge fixed Maxwell equations. After a separation of variables, the radial mode equations form a complicated system of coupled linear ODEs. We outline a precise abstract strategy to decouple this system into triangular form, where the diagonal blocks consist of spin-s scalar Regge-Wheeler equations, with $s = 0$ or 1. This strategy is then implemented to give an explicit transformation of the radial mode equations (with nonzero frequency and angular momentum) into this triangular form. Our decoupling goes a step further than previous results in the literature by making the triangular form explicit and reducing it as much as possible. Also, with the help of our abstractly formulated decoupling strategy, we have significantly streamlined both the presentation of the final results and the intermediate calculations. Finally, we note that the vector wave equation is a simple model for more complicated equations, like harmonic (or de Donder) gauge fixed linearized gravity, and backgrounds, like Kerr, where we expect the same abstract decoupling strategy to work as well.

1. Introduction

The study of linear wave-like equations on a Schwarzschild black hole background has a long history and many applications [1, 2], both for scalar fields as well as higher rank tensor fields. In this work, we will concentrate on the vector wave equation. This equation, besides its purely mathematical significance as a model for more general tensor wave equations, can be seen as the harmonic (or Lorenz) gauge-fixed version of Maxwell equations [3, Sec.15.1], [4], or also as the residual gauge equation for harmonic (or de Donder) gauge-fixed linearized gravity [5, 6, 7].

While the vector wave equation admits a complete separation of variables on Schwarzschild, the resulting radial mode equations still represent a complicated system of coupled ordinary differential equations (ODEs) in the radial variable. What is worse is that, though this ODE system is formally self-adjoint (it can be put into matrix Sturm-Liouville form), it naturally defines a symmetric unbounded operator only on a Krein space (a topological vector space with an indefinite scalar product). This indefiniteness is directly traced to the indefiniteness of the Lorentzian background metric and does not occur in the analogous Riemannian problem. Thus, the abstract spectral theory of symmetric and self-adjoint operators on a Hilbert space is not applicable in this case, which significantly complicates important and fundamental questions about this ODE system. For example: Can we prove or disproved that its frequency spectrum is

IOP Conf. Series: Journal of Physics: Conf. Series **968** (2018) 012006 doi:10.1088/1742-6596/968/1/012006

purely real? Can we prove that its generalized eigenfunctions are complete? Can we construct a spectral representation for the radial Green function?

To be able to answer these questions, one must appeal to some very special structural properties of the vector wave equation. Namely, in this work, we will show that the radial mode equations (at least for modes with non-zero frequency and angular momentum) can be decoupled into a triangular system, where the diagonal blocks are the spin-s Regge-Wheeler equations, with $s = 0$ or 1. These Regge-Wheeler equations are then of standard scalar Sturm-Liouville form, with very well understood spectral properties. Such a decoupling has been previously discussed in the literature in [7] and [4], though not in as explicit and conceptually clear terms as we present below. In fact, the full explicit details of how the original mode equations transform into the decoupled form and back are not easy to extract from these references. The strategy that we follow is basically the one of [7].[1] Though, our results go a step further, by reducing the resulting triangular form as much as possible.

In principle, starting from the triangular decoupled form, the spectral properties of the Regge-Wheeler equations can completely determine the spectral properties of the original radial mode equations. However, the details of such a spectral analysis are left for future works.

In Section 2 we fix our notation, introduce the formal properties of differential equations and operators, and formulate our abstract decoupling strategy. In Section 3 we present a complete separation of variables for the vector wave equation on Schwarzschild spacetime and apply our strategy to give an explicit triangular decoupling of the resulting radial mode equations into spin-0 and spin-1 Regge-Wheeler scalars. In Section 4, we discuss in more detail the novel aspects of our results, as well as several avenues of further investigation.

2. Formal properties of differential equations and operators

2.1. Morphisms and cochain maps between differential equations

For the purposes of this work, a *differential operator*, say f, is always linear, with smooth coefficients, which we will write as $f[u]$, where u is a possibly vector valued function. Thus f can always be thought of as a matrix of scalar differential operators acting on the components of u. By a *scalar differential operator*, we mean an operator that takes possibly vector valued functions into scalar valued functions (corresponding to a matrix with a single row). Scalars are real or complex numbers (for the sake of generality we will generally allow complex scalars). Differential operators could be of any order, including order zero, which just corresponds to multiplication by some matrix valued function. While the operators can be thought of as acting on smooth (vector valued) functions, we will be mostly concerned with composition identities among operators, and so we will not bother specifying precisely the domain or codomain of each operator. This information can always be deduced from the context. Also, for the abstract discussion below, it is also not necessary to fix the number of independent variables, but we will be mostly concerned with applications to ordinary differential operators (i.e., acting on functions of a single independent variable).

Below, we state some basic definitions and results concerning differential equations and morphisms between them, which essentially correspond to differential operators that map solutions to solutions. The presentation is logically self-contained and does not extend far beyond what is needed in the rest of the paper. However, for proper context, these ideas can be seen as simple special cases of the more general frameworks of D-modules [8, Sec.10.5], the category of differential equations [9, Sec.VII.5] or homological algebra [10]. We will not delve

[1] It is by no means a simple task to extract this strategy from [7], as it only becomes apparent as the common pattern in the detailed and explicit calculations done for a sequence of examples of increasing complexity. Still, that work should be credited as (to our knowledge) the first to carry out this strategy explicitly for the example of the vector wave equation as well as the significantly more complicated Lichnerowicz equation, the latter corresponding to harmonic gauge fixed linearized gravity, on Schwarzschild.

IOP Conf. Series: Journal of Physics: Conf. Series **968** (2018) 012006 doi:10.1088/1742-6596/968/1/012006

into the precise connection with these larger frameworks, but it is useful to mention that all arrows/morphisms need to be reversed when our statements are interpreted in the language of D-modules.

When referring to a morphism between differential equations, we essentially mean a differential operator that maps solutions to solutions. Sometimes given a differential equation $e[u] = 0$ we will also refer to the operator e as the equation. This should not lead to any confusion.

Definition 2.1. *Consider two differential equations $e[u] = 0$ and $\bar{e}[\bar{u}] = 0$.*

(a) A differential operator $\bar{u} = k[u]$ is a morphism *from $e[u] = 0$ to $\bar{e}[\bar{u}] = 0$ when for any scalar differential operator of the form $\bar{p} \circ \bar{e}[\bar{u}]$ there exists a scalar differential operator of the form $q \circ e[u]$ such that*

$$\bar{p} \circ \bar{e}[k[u]] = q \circ e[u], \tag{1}$$

and hence the map $p \mapsto k_p$ given by the identity

$$p \circ k = k_p - h_p \circ e \tag{2}$$

is well-defined on equivalence classes modulo $(\cdots) \circ \bar{e}$ and $(\cdots) \circ e$ respectively.

(b) A pair of differential operators $\bar{u} = k[u]$ and $\bar{v} = g[v]$ is a cochain map *from $e[u] = 0$ to $\bar{e}[\bar{u}] = 0$ when they satisfy the identity $\bar{e} \circ k = g \circ \bar{e}$.*

It is easy to check that a *cochain map* is always a *morphism* and, as we will see shortly, a *morphism* always gives rise to a *cochain map*, but not necessarily uniquely. The two definitions are complementary to each other. Since a cochain map comes with more structure, it is often more convenient for algebraic manipulations. On the other hand, some properties of morphisms may be easier to check. We illustrate morphisms and cochain maps respectively as

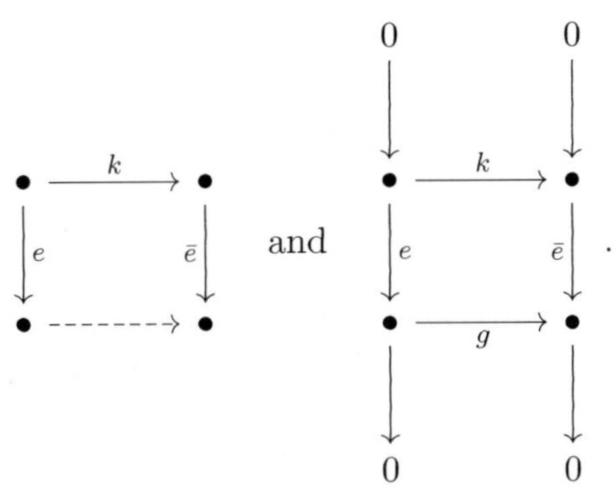

(3)

The second diagram is meant to illustrate the terminology *cochain map*, which comes from homological algebra [10]. There, the primary objects are *cochain complexes*, which are sequences of linear spaces (or more generally modules, or even just abelian groups) connected by linear maps, of which any two successive ones compose to zero. What is of interest about a cochain complex is its *cohomology* (the quotient of the kernel of one linear map by the image of the preceding one). In our case the property of being a cochain complex is trivially satisfied by the columns of the second diagram and the cohomologies correspond to the spaces $\ker e$ and $\operatorname{coker} e$ of our differential operator e. Morphisms between cochain complexes are sequences of maps that commute with the linear maps of each complex. In our case, this property is illustrated by the commutativity (the identity $\bar{e} \circ k = g \circ e$) of the middle square of the second diagram. It is straight forward to check that cochain maps induce well-defined maps on cohomology, which in our case corresponds to k mapping solutions of $e[u] = 0$ ($u \in \ker e$) to solutions of $\bar{e}[\bar{u}] = 0$

($u \in \ker \bar{e}$). In the sequel, we will always be concerned with just the properties of the middle square of the second diagram, hence the zero maps (illustrated above and below the middle square) will from now on be omitted and left implicit.

Lemma 2.1. *A morphism k from $e[u] = 0$ to $\bar{e}[\bar{u}] = 0$ always induces a cochain map k, g.*

Proof. Choosing $\bar{p} = \mathrm{id}$ in the definition of a morphism (Definition 2.1(a)), implies that there exists an operator g such that $\bar{e} \circ k = g \circ e$. The last identity also implies that k, g is a cochain map (Definition 2.1(b)). □

Of course in general the operator g completing the morphism k to a cochain map may not be unique. The next two definitions introduce some parallel properties of morphisms and cochain maps.

Definition 2.2. *A morphism k from $e[u] = 0$ to $\bar{e}[\bar{u}] = 0$ is called* on-shell injective (surjective) *if the map $p \mapsto k_p$ is surjective (injective) from equivalence classes modulo $(\cdots) \circ \bar{e}$ to equivalence classes modulo $(\cdots) \circ e$. We say* on-shell bijective *to mean both on-shell injective and surjective. We say that k is* on-shell vanishing *when $p \circ k$ belongs to the same equivalence class as 0 modulo $(\cdots) \circ e$ for any operator p.*

Definition 2.3. *A cochain map k, g from $e[u] = 0$ to $\bar{e}[\bar{u}] = 0$ is said to be* induced by a homotopy *if $k = h \circ e$ and $g = \bar{e} \circ h$ for some operator h referred to as a* (cochain) homotopy*:*

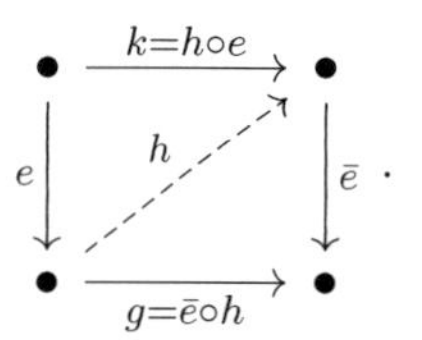

(4)

We say that k, g has a left (right) inverse up to homotopy *if there exists a cochain map $\bar{k}, \bar{g}$ from $\bar{e}[\bar{u}] = 0$ to $e[u] = 0$ such that*

$$\begin{aligned} \bar{k} \circ k &= \mathrm{id} - h \circ e \\ \bar{g} \circ g &= \mathrm{id} - e \circ h \end{aligned} \qquad \begin{pmatrix} k \circ \bar{k} = \mathrm{id} - \bar{h} \circ \bar{e} \\ g \circ \bar{g} = \mathrm{id} - \bar{e} \circ \bar{h} \end{pmatrix}$$

for some homotopy h of $e[u] = 0$ ($\bar{h}$ of $\bar{e}[\bar{u}] = 0$) into itself. A cochain map k, g that has a two-sided inverse up to homotopy is called an equivalence up to homotopy.

The last two sets of definitions are related but not exactly equivalent. In general, the existence of a left or right inverse is a stronger property than being on-shell injective or surjective. Though, under certain hypotheses, the link could be made stronger.

Definition 2.4. *Consider a differential equation $e[u] = 0$. Any operator p such that $e \circ p = 0$ is called a* gauge symmetry (generator) *of e. Any operator q such that $q \circ e = 0$ is called a* Noether identity *(or* compatibility operator*) of e. The equation is said to be* under-determined *if it has a non-zero gauge symmetry and* over-determined *if it has a non-zero Noether identity. The equation is said to be* over-under-determined *if it has both non-zero gauge symmetries and Noether identities and* determined *(or* normal*) if it has neither.*

Lemma 2.2. *Consider equations $e[u] = 0$ and $\bar{e}[\bar{u}] = 0$.*

(a) When e is not over-determined, a morphism $\bar{u} = k[u]$ induces a unique *cochain map k, g.*

(b) A cochain map k, g has a left (right) inverse up to homotopy $\Longrightarrow$ k is on-shell injective (surjective). If k, g is induced by a homotopy, then k is on-shell vanishing.

IOP Conf. Series: Journal of Physics: Conf. Series **968** (2018) 012006 doi:10.1088/1742-6596/968/1/012006

(c) When a morphism $\bar{u} = k[u]$ is on-shell bijective, it has a two-sided inverse morphism $u = \bar{k}[\bar{u}]$.

(d) When neither e nor $\bar{e}$ is over-determined, an on-shell bijective morphism $\bar{u} = k[u]$ induces an equivalence up to homotopy. More precisely, it induces a cochain map k, g (with g unique) and a two-sided inverse up to homotopy $\bar{k}, \bar{g}$.

Proof. (a) From Lemma 2.1, we already know that there exists at least one operator g such that k, g is a cochain map. Let g' be another such operator. Then $(g' - g) \circ e = \bar{e} \circ (k - k) = 0$. Hence, if e has no non-vanishing Noether identities (Definition 2.4), we must have $g' = g$.

(b) Suppose that $\bar{k}, \bar{g}$ is a left inverse to k, g up to homotopy. Then, for any operator $q[u]$, the operator $\bar{p} = q \circ \bar{k}$ satisfies

$$\bar{p} \circ k = q \circ (\bar{k} \circ k) = q - (q \circ h) \circ e\,. \tag{5}$$

Hence k is on-shell injective.
On the other hand, suppose that $\bar{k}, \bar{g}$ is a right inverse to k, g up to homotopy. Then any operator $\bar{p}$ such that $\bar{p} \circ k = q \circ e$, we have the identity

$$p \circ (\mathrm{id} - \bar{h} \circ \bar{e}) = (p \circ k) \circ \bar{k} = q \circ (e \circ \bar{k}) = (q \circ \bar{g}) \circ \bar{e}\,. \tag{6}$$

Hence we must have $\bar{p} = (q \circ \bar{g} + p \circ h) \circ \bar{e}$ and therefore k is on-shell surjective.
Finally if $k = h \circ e$ and $g = \bar{e} \circ h$, then for any operator p we have $p \circ k = (p \circ h) \circ e$, meaning that k is on-shell vanishing.

(c) By on-shell injectivity, there must exist operators $\bar{k}$ and h such that $\bar{k} \circ k = \mathrm{id} - h \circ e$. Then

$$(\mathrm{id} - k \circ \bar{k}) \circ k = k \circ (\mathrm{id} - \bar{k} \circ k) = (k \circ h) \circ e\,. \tag{7}$$

Hence, by on-shell surjectivity, there must exist an operator $\bar{h}$ such that $k \circ \bar{k} = \mathrm{id} - \bar{h} \circ \bar{e}$. Now, it remains only to check that $\bar{k}$ is actually a morphism. For that, note the identity

$$(e \circ \bar{k}) \circ k = e \circ (\bar{k} \circ k) = e - e \circ h \circ e = (\mathrm{id} - e \circ h) \circ e\,. \tag{8}$$

But by on-shell surjectivity of k this means that there exists an operator $\bar{g}$ such that $e \circ \bar{k} = \bar{g} \circ \bar{e}$, meaning that $\bar{k}, \bar{g}$ is a cochain map and hence $\bar{k}$ itself is a morphism that is a two-sided on-shell inverse to k.

(d) By part (c), we know that an on-shell bijective morphism k has a two sided inverse morphism $\bar{k}$, satisfying

$$\bar{k} \circ k = \mathrm{id} - h \circ e \quad \text{and} \quad k \circ \bar{k} = \mathrm{id} - \bar{h} \circ \bar{e}\,, \tag{9}$$

for some operators h and $\bar{h}$. By part (a), they can both be uniquely completed to cochain maps, k, g and $\bar{k}, \bar{g}$. It remains to check the relation between g and $\bar{g}$. Consider the identities

$$(\bar{g} \circ g) \circ e = \bar{g} \circ (g \circ e) = (\bar{g} \circ \bar{e}) \circ k = e \circ (\bar{k} \circ k) = (\mathrm{id} - e \circ h) \circ e\,, \tag{10}$$

$$(g \circ \bar{g}) \circ \bar{e} = g \circ (\bar{g} \circ \bar{e}) = (g \circ e) \circ \bar{k} = \bar{e} \circ (k \circ \bar{k}) = (\mathrm{id} - \bar{e} \circ \bar{h}) \circ \bar{e}\,. \tag{11}$$

Since neither e nor $\bar{e}$ has non-vanishing Noether identities, we must conclude that

$$\bar{g} \circ g = \mathrm{id} - e \circ h \quad \text{and} \quad g \circ \bar{g} = \mathrm{id} - \bar{e} \circ \bar{h}\,. \tag{12}$$

Hence, the cochain map $\bar{k}, \bar{g}$ is a two-sided to k, g up to homotopy.

□

IOP Conf. Series: Journal of Physics: Conf. Series **968** (2018) 012006 doi:10.1088/1742-6596/968/1/012006

2.2. Block-triangular decoupling

In this section, we show some equations can be transformed (or *decoupled*) into block triangular form. These results will be the building blocks of our abstract decoupling strategy, which will be discussed in Section 2.5. Note that we do not state these results in their most general form, but only in sufficient generality to be useful later on.

Lemma 2.3. *Consider the differential equations* $e[u] = 0$ *and* $f[v] = 0$. *A morphism* $v = k[u]$ *induces an on-shell bijective morphism from* $e[u] = 0$ *to* $\bar{e}[\bar{u}] = 0$, *with* $\bar{u} = \left[\begin{smallmatrix} u \\ v \end{smallmatrix}\right]$ *and* $\bar{e}$ *in upper block-triangular form, namely*

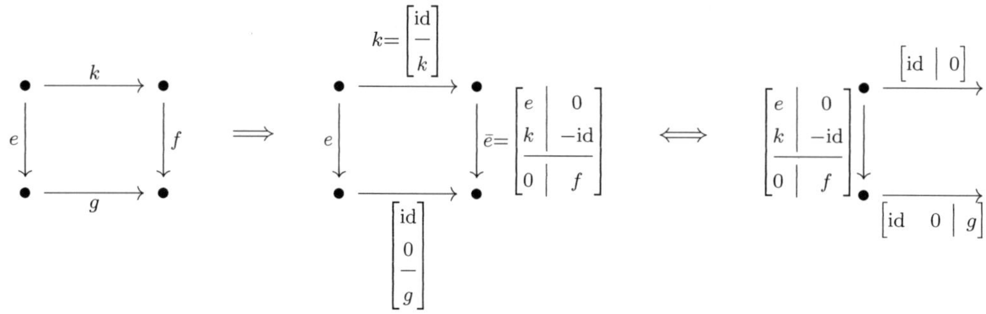

(13)

where the last two morphisms are mutually on-shell inverse.

We say that the morphism k *decouples* f from e. Once f has been decoupled into a lower diagonal block, any transformations of remaining upper diagonal block transform the block-triangular form as follows.

Lemma 2.4. *Consider the equations* $e[u] = 0$, $\bar{e}[\bar{u}] = 0$, $f[v] = 0$, *and*

$$\begin{bmatrix} e & \Delta \\ 0 & f \end{bmatrix} \begin{bmatrix} u \\ v \end{bmatrix} = 0\,, \tag{14}$$

where the operator Δ *has the property that for any Noether identity* $n \circ e = 0$ *we have* $n \circ \Delta = m \circ f$ *for some operator* m. *An on-shell bijective morphism* k *between* e *and* $\bar{e}$, *which more precisely satisfies the identities (which are slightly weaker than equivalence up to homotopy)*

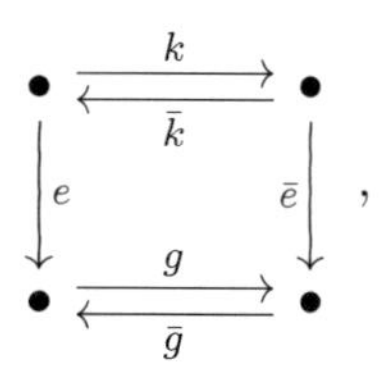

$$\bar{k} \circ k = \mathrm{id} - h \circ e\,, \qquad \bar{g} \circ g = \mathrm{id} - e \circ h - n\,. \tag{15}$$

with a Noether identity $n \circ e = 0$, *induces the following mutually on-shell bijective morphisms of equations in triangular form, where* $\bar{\Delta} = g \circ \Delta$:

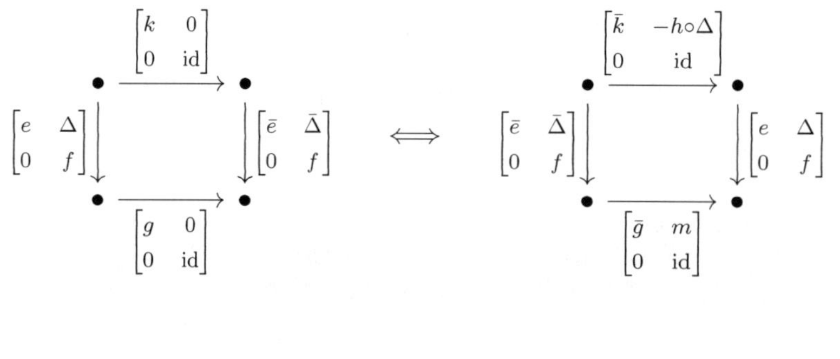

(16)

IOP Conf. Series: Journal of Physics: Conf. Series **968** (2018) 012006 doi:10.1088/1742-6596/968/1/012006

Of course, we should note that the block-triangular equations obtained in the earlier Lemma 2.3 satisfy the hypotheses of Lemma 2.4.

Proof. Consider the following identities:

$$\begin{bmatrix} \bar{e} & \bar{\Delta} \\ 0 & f \end{bmatrix} \circ \begin{bmatrix} k & 0 \\ 0 & \mathrm{id} \end{bmatrix} - \begin{bmatrix} g & 0 \\ 0 & \mathrm{id} \end{bmatrix} \circ \begin{bmatrix} e & \Delta \\ 0 & f \end{bmatrix} = \begin{bmatrix} \bar{e} \circ k - g \circ e & \bar{\Delta} - g \circ \Delta \\ 0 & f - f \end{bmatrix}, \tag{17}$$

$$\begin{aligned} &\begin{bmatrix} e & \Delta \\ 0 & f \end{bmatrix} \circ \begin{bmatrix} \bar{k} & -h \circ \Delta \\ 0 & \mathrm{id} \end{bmatrix} - \begin{bmatrix} \bar{g} & m \\ 0 & \mathrm{id} \end{bmatrix} \circ \begin{bmatrix} \bar{e} & g \circ \Delta \\ 0 & f \end{bmatrix} \\ &\qquad = \begin{bmatrix} e \circ \bar{k} - \bar{g} \circ \bar{e} & (\mathrm{id} - e \circ h - n - \bar{g} \circ g) \circ \Delta + (n \circ \Delta - m \circ f) \\ 0 & f - f \end{bmatrix}. \end{aligned} \tag{18}$$

They show that the diagrams in the Lemma show actual cochain maps. It is also straight forward to see that they are mutually inverse up to homotopy. □

We also have the analogous

Lemma 2.5. *Consider equations $e[u] = 0$, $\bar{e}[\bar{u}] = 0$, $f[v] = 0$ and*

$$\begin{bmatrix} e & \Delta \\ 0 & f \end{bmatrix} \begin{bmatrix} u \\ v \end{bmatrix} = 0, \tag{19}$$

An equivalence up to homotopy k, g between e and $\bar{e}$

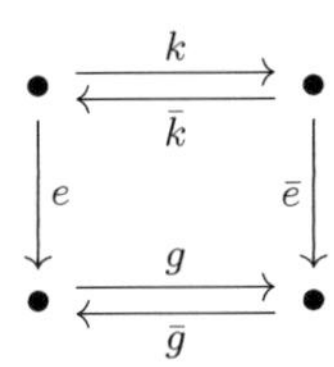

$$\bar{k} \circ k = \mathrm{id} - h \circ e, \qquad \bar{g} \circ g = \mathrm{id} - e \circ h. \tag{20}$$

induces the following mutually on-shell bijective morphisms of equations in triangular form, where $\Delta = \bar{\Delta} \circ k$:

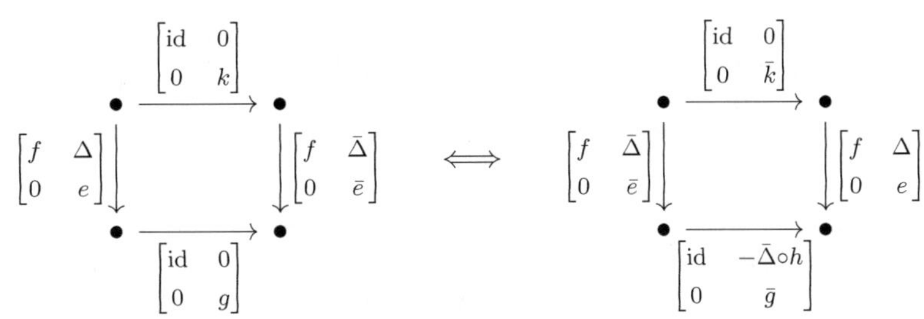

(21)

Proof. Consider the following identities:

$$\begin{bmatrix} f & \bar{\Delta} \\ 0 & \bar{e} \end{bmatrix} \circ \begin{bmatrix} \mathrm{id} & 0 \\ 0 & k \end{bmatrix} - \begin{bmatrix} \mathrm{id} & 0 \\ 0 & g \end{bmatrix} \circ \begin{bmatrix} f & \Delta \\ 0 & e \end{bmatrix} = \begin{bmatrix} f - f & \bar{\Delta} \circ k - \Delta \\ 0 & \bar{e} \circ k - g \circ e \end{bmatrix}, \tag{22}$$

$$\begin{bmatrix} f & \Delta \\ 0 & e \end{bmatrix} \circ \begin{bmatrix} \mathrm{id} & 0 \\ 0 & \bar{k} \end{bmatrix} - \begin{bmatrix} \mathrm{id} & -\bar{\Delta} \circ h \\ 0 & \bar{g} \end{bmatrix} \circ \begin{bmatrix} f & \bar{\Delta} \\ 0 & \bar{e} \end{bmatrix} = \begin{bmatrix} f - f & \bar{\Delta} \circ (k \circ \bar{k} - \mathrm{id} + \bar{h} \circ \bar{e}) \\ 0 & e \circ \bar{k} - \bar{g} \circ \bar{e} \end{bmatrix}. \tag{23}$$

They show that the diagrams in the Lemma show actual cochain maps. It is also straight forward to see that they are mutually inverse up to homotopy. □

2.3. Reducing the triangular form
Given an equation that is in block-*triangular* form, when is it equivalent to an equation in block-*diagonal* form? That is, when can an equation in triangular form be further reduced?

In the sequel, we will only need the following result.

Lemma 2.6. *Consider the mutually inverse up to homotopy cochain maps*

$$\begin{array}{ccc} \bullet & \xrightarrow{\begin{bmatrix} \mathrm{id} & \delta \\ 0 & \mathrm{id} \end{bmatrix}} & \bullet \\ {\scriptstyle \begin{bmatrix} e & f \\ 0 & \bar{e} \end{bmatrix}} \Big\downarrow & & \Big\downarrow {\scriptstyle \begin{bmatrix} e & 0 \\ 0 & \bar{e} \end{bmatrix}} \\ \bullet & \xrightarrow[\begin{bmatrix} \mathrm{id} & \varepsilon \\ 0 & \mathrm{id} \end{bmatrix}]{} & \bullet \end{array} \quad \Longleftrightarrow \quad \begin{array}{ccc} \bullet & \xrightarrow{\begin{bmatrix} \mathrm{id} & -\delta \\ 0 & \mathrm{id} \end{bmatrix}} & \bullet \\ {\scriptstyle \begin{bmatrix} e & 0 \\ 0 & \bar{e} \end{bmatrix}} \Big\downarrow & & \Big\downarrow {\scriptstyle \begin{bmatrix} e & f \\ 0 & \bar{e} \end{bmatrix}} \\ \bullet & \xrightarrow[\begin{bmatrix} \mathrm{id} & -\varepsilon \\ 0 & \mathrm{id} \end{bmatrix}]{} & \bullet \end{array} . \tag{24}$$

The two squares in these diagrams actually commute (that is, we can kill the off-diagonal block f) if and only if the operators δ and ε satisfy the identity

$$e \circ \delta = f + \varepsilon \circ \bar{e} . \tag{25}$$

Essentially, the operators δ and ε allow us to solve the equation is $e[u] = f[\bar{u}]$ by $u = \delta[\bar{u}]$ whenever $\bar{e}[\bar{u}] = 0$.

Proof. The desired result follows immediately from the identity

$$\begin{bmatrix} e & 0 \\ 0 & \bar{e} \end{bmatrix} \circ \begin{bmatrix} \mathrm{id} & \delta \\ 0 & \mathrm{id} \end{bmatrix} - \begin{bmatrix} \mathrm{id} & \varepsilon \\ 0 & \mathrm{id} \end{bmatrix} \circ \begin{bmatrix} e & f \\ 0 & \bar{e} \end{bmatrix} = \begin{bmatrix} e - e & e \circ \delta - f - \varepsilon \circ \bar{e} \\ 0 & \bar{e} - \bar{e} \end{bmatrix} . \tag{26}$$

The reverse direction is completely analogous. □

2.4. Formal adjoints
Recall that we are only dealing with linear ordinary differential operators, with derivative operator ∂_r, where each operator can be written as a matrix of scalar differential operators with smooth coefficients. Given a matrix linear differential operator k, we define its formal adjoint k^* using the following rules. If k is a zero-th order scalar operator acting on a single scalar argument, basically multiplication by some scalar function $k(r)$, then k^* is multiplication by the complex conjugate function $k(r)^*$. If $k^* = \partial_r$ acting on a single scalar argument, then $k^* = -\partial_r$. If k and g are scalar operators acting on a single scalar argument, then $(k + g)^* = k^* + g^*$ and $(k \circ g)^* = g^* \circ k^*$. Finally, if $k = [k_{ij}]$ is a matrix of scalar differential operators, then $(k^*)_{ji} = k^*_{ji}$. It is straight forward to check that this gives rise to a consistent and unique definition, which satisfies $(k \circ g)^* = g^* \circ k^*$ and $k^{**} = k$, now with any composable operators k and g, and agrees with the common notion of formal adjoint. When dealing with a family of operators k_ω parametrized by a possibly complex parameter ω, we adopt the convension that the family of adjoint operators is parametrized as $(k^*)_\omega = (k_{\omega^*})^*$. With this convention, $(k^*)_\omega = (k_\omega)^*$ for real values $\omega = \omega^*$ of the parameter. An operator satisfying $k^* = k$ is called *(formally) self-adjoint*.

Sometimes, it will be convenient for us, when introducing a new differential operator, to write it directly as k^*, even if we have not explicitly introduced the operator k. This is possible, since any differential operator is the formal adjoint of some differential operator. This convention allows us to highlight some special features of the equations and differential operators that we will be working with, and also reigns in a bit the proliferation of new notation.

2.5. *Gauge fixing, gauge-invariant fields and triangular decoupling*

In this section, we give an abstract overview of the triangular decoupling procedure that will be illustrated on a specific example in Section 3. The abstract approach allows us to separate the general strategy from the calculational details of a specific example, so that it could be applied in other examples as well. One benefit of having a general strategy is that it eliminates much trial-and-error and unnecessary calculations in more complex examples.

We start with a system of differential equations $E[u] = 0$ and aim to find explicit equivalence of this system to one that is in block upper-triangular form, with as many off-diagonal terms set to zero (or *reduced*) as possible. Such a *triangular decoupling* will be possible because of the large amount of structure that we can assume about our equation. Essentially, we expect E to be the gauge-fixed form of a more fundamental self-adjoint equation $E_0[u] = 0$ that has both gauge symmetries and (hence necessarily) Noether identities. The gauge fixing and the modifications of E from E_0 are chosen such that E remains self-adjoint and it induces a nice residual gauge equation. We will refer to such a gauge fixing condition as *harmonic*, because harmonic or generalized harmonic gauges tend to satisfy all of these requirements.

While we do try to keep track of self-adjointness of our equations, which is helpful at least as an error-check on explicit calculations, adjoint operators do not play a specific role in the decoupling strategy outlined below. Hence, it would work just as well if the operators that appear as adjoints are replaced by independent operators, which satisfy all the same composition identities.

The general strategy will be to show that the original dependent variables separate into *gauge modes*, *gauge invariant modes* and *constraint violating modes*, which are coupled to each other in a hierarchical (hence *triangular*) and minimal way.

Let us introduce the following differential operators and identities among them:

- E_0—self-adjoint equation with gauge symmetry,
- D—gauge symmetry generator, $E_0 \circ D = 0$,
- D^*—Noether identity, $D^* \circ E_0 = 0$,
- T—harmonic gauge fixing condition, $T \circ D = R \circ \mathcal{D}_D$,
- $\mathcal{D}_D$—self-adjoint residual gauge equation,
- R—self-adjoint gauge fixing correction, $R^* = R$,
- $E = E_0 + T^* \circ \bar{R} \circ T$—self-adjoint gauge-fixed equation, $\bar{R} = R^{-1}$,
- Φ_0—gauge invariant field, $\Phi_0 \circ D = 0$,
- $\mathcal{D}_\Phi$—self-adjoint dynamical equation for gauge-invariant fields, $\mathcal{D}_\Phi \circ \Phi_0 = \bar{\Phi}_0^* \circ E_0$, with $E \circ D = T^* \circ \mathcal{D}_D$ and $\mathcal{D}_D^* \circ T = D^* \circ E$.

The operators

$$D[\phi_0]\,, \quad \Phi_0\,, \quad T \tag{27}$$

respectively separate out, in a sense that will become clear by the end of this section, the *gauge modes*, the *gauge invariant modes* and the *constraint violating modes.* The general strategy will be to show that these modes couple to each other only in a hierarchical (hence *triangular*) way, and to find slight modifications to them that reduce this coupling as much as possible. The subscript s on ϕ_s and ψ_s indicates the spin of the corresponding Regge-Wheeler operators $\mathcal{D}_s$ that appear on the diagonal of the decoupled form.

The above operators fit into the following commutative diagrams:

$$\begin{CD} \bullet @>D>> \bullet \\ @V{\mathcal{D}_D}VV @VV{E}V \\ \bullet @>>{T^*}> \bullet \end{CD}\,, \qquad \begin{CD} \bullet @>T>> \bullet \\ @V{E}VV @VV{\mathcal{D}_D^*}V \\ \bullet @>>{D^*}> \bullet \end{CD}\,, \qquad \begin{CD} \bullet @>{\Phi_0}>> \bullet \\ @V{E_0}VV @VV{\mathcal{D}_\Phi}V \\ \bullet @>>{\bar{\Phi}_0^*}> \bullet \end{CD}\,. \tag{28}$$

IOP Conf. Series: Journal of Physics: Conf. Series **968** (2018) 012006 doi:10.1088/1742-6596/968/1/012006

Even though we have $\mathcal{D}_D^* = \mathcal{D}_D$, we use this notation to formally distinguish between the dynamical equation satisfied by the longitudinal/constraint-violating modes T and the residual gauge modes D. Note that the operator $\bar{\Phi}_0^*$ is not uniquely fixed by completing the commutative diagram between E_0, Φ_0 and $\mathcal{D}_\Phi$. If $\bar{\Phi}_1^* = \bar{\Phi}_0^* + N$ is any other such operator, then $N \circ E_0 = 0$ and hence N is a Noether identity for E_0.

As a first step, we use the operators T and D^* and Lemma 2.3 to decouple $\mathcal{D}_D^*$ from E. We have the following morphism from E to a triangular form:

$$\begin{CD} \bullet @>{\begin{bmatrix}\mathrm{id}\\ T\end{bmatrix}}>> \bullet \\ @V{E}VV @VV{\begin{bmatrix}E & 0\\ T & -\mathrm{id}\\ 0 & \mathcal{D}_D^*\end{bmatrix}}V \\ \bullet @>>{\begin{bmatrix}\mathrm{id}\\ 0\\ D^*\end{bmatrix}}> \bullet \end{CD} \quad\Longleftrightarrow\quad \begin{CD} \bullet @>{\begin{bmatrix}\mathrm{id} & 0\end{bmatrix}}>> \bullet \\ @V{\begin{bmatrix}E & 0\\ T & -\mathrm{id}\\ 0 & \mathcal{D}_D^*\end{bmatrix}}VV @VV{E}V \\ \bullet @>>{\begin{bmatrix}\mathrm{id} & 0 & 0\end{bmatrix}}> \bullet \end{CD}\,, \tag{29}$$

where the last two morphisms are on-shell mutually inverse.

Next, we would like to use the operators Φ_0 and $\bar{\Phi}_0^*$ to decouple $\mathcal{D}_\Phi$ from $[\begin{smallmatrix}E\\T\end{smallmatrix}]$. Let us introduce the operator

$$\bar{\Delta}_\Phi = \bar{\Phi}_0^* \circ T^* \circ \bar{R}\,, \tag{30}$$

which is necessary to decouple $\mathcal{D}_\Phi$, since it completes the commutative diagram

$$\begin{CD} \bullet @>{\Phi_0}>> \bullet \\ @V{\begin{bmatrix}E\\ T\end{bmatrix}}VV @VV{\mathcal{D}_\Phi}V \\ \bullet @>>{\begin{bmatrix}\bar{\Phi}_0^* & -\bar{\Delta}_\Phi\end{bmatrix}}> \bullet \end{CD}\,. \tag{31}$$

Once again, applying Lemma 2.3 we get the mutually on-shell inverse morphisms

$$\begin{CD} \bullet @>{\begin{bmatrix}\mathrm{id}\\ \Phi_0\end{bmatrix}}>> \bullet \\ @V{\begin{bmatrix}E\\ T\end{bmatrix}}VV @VV{\begin{bmatrix}E & 0\\ T & 0\\ \Phi_0 & -\mathrm{id}\\ 0 & \mathcal{D}_\Phi\end{bmatrix}}V \\ \bullet @>>{\begin{bmatrix}\mathrm{id} & 0\\ 0 & \mathrm{id}\\ 0 & 0\\ \bar{\Phi}_0^* & -\bar{\Delta}_\Phi\end{bmatrix}}> \bullet \end{CD} \quad\Longleftrightarrow\quad \begin{CD} \bullet @>{\begin{bmatrix}\mathrm{id} & 0\end{bmatrix}}>> \bullet \\ @V{\begin{bmatrix}E & 0\\ T & 0\\ \Phi_0 & -\mathrm{id}\\ 0 & \mathcal{D}_\Phi\end{bmatrix}}VV @VV{\begin{bmatrix}E\\ T\end{bmatrix}}V \\ \bullet @>>{\begin{bmatrix}\mathrm{id} & 0 & 0 & 0\\ 0 & \mathrm{id} & 0 & 0\end{bmatrix}}> \bullet \end{CD}\,. \tag{32}$$

The operator $\bar{\Delta}_\Phi$ is an obstacle to decoupling $\mathcal{D}_\Phi$ directly from E. This obstacle becomes trivial when $\bar{\Delta}_\Phi = \mathcal{D}_\Phi \circ q$ for some operator q, since then replacing Φ_0 with $\Phi_0 + q^* \circ T$ and setting $\bar{\Delta}_\Phi = 0$ does the required job. If we had used $\Phi_0 + q \circ T$ from the start, then the final triangular form that we will arrive at will not have any coupling between T and $\Phi_0 + q \circ T$. We will actually be able to achieve such a reduced triangular form, by using the strategy from Section 2.3 at the very end.

IOP Conf. Series: Journal of Physics: Conf. Series **968** (2018) 012006 doi:10.1088/1742-6596/968/1/012006

Note that due to the identities involving D, T, Φ_0 and E, the operator D remains a morphism from $\mathcal{D}_D$ to each of the E, E-T and E-T-Φ_0 equations. We will proceed under the hypothesis that in that last case *the morphism D is actually on-shell bijective.* That is, by an application of Lemmas 2.2 and 2.1 we have the following mutually on-shell inverse morphisms

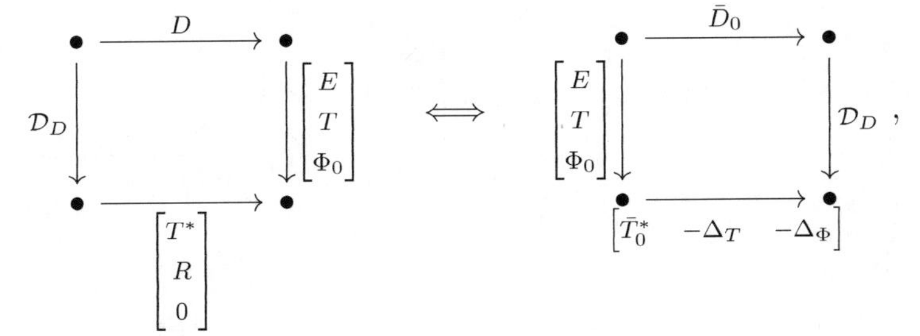

(33)

for some operators $\bar{T}_0^*$, Δ_T and Δ_Φ. It would be convenient if the above diagrams actually described an equivalence up to homotopy. In general that will not be true, but only due to the rather restricted notion of *up to homotopy* that we have adopted in this work. On the other hand, we make another hypothesis that the inverse relationship between the above cochain maps takes on the following slightly more general form, which we have actually already seen in the hypotheses of Lemma 2.4:

$$\bar{D}_0 \circ D = \mathrm{id} - h_D \circ \mathcal{D}_D\,, \tag{34}$$

$$\begin{bmatrix} \bar{T}_0^* & -\Delta_T & -\Delta_\Phi \end{bmatrix} \circ \begin{bmatrix} T^* \\ R \\ 0 \end{bmatrix} = \mathrm{id} - \mathcal{D}_D \circ h_D\,, \tag{35}$$

$$D \circ \bar{D}_0 = \mathrm{id} - \begin{bmatrix} \bar{h}_E & \bar{T}_2 & \bar{\Phi}_2 \end{bmatrix} \circ \begin{bmatrix} E \\ T \\ \Phi_0 \end{bmatrix}\,, \tag{36}$$

$$\begin{bmatrix} T^* \\ R \\ 0 \end{bmatrix} \circ \begin{bmatrix} \bar{T}_0^* & -\Delta_T & -\Delta_\Phi \end{bmatrix} = \begin{bmatrix} \mathrm{id} & 0 & 0 \\ 0 & \mathrm{id} & 0 \\ 0 & 0 & \mathrm{id} \end{bmatrix} - \begin{bmatrix} E \\ T \\ \Phi_0 \end{bmatrix} \circ \begin{bmatrix} \bar{h}_E & \bar{T}_2 & \bar{\Phi}_2 \end{bmatrix} - \bar{n}\,, \tag{37}$$

$$\text{where} \quad \bar{n} \circ \begin{bmatrix} 0 & 0 \\ 0 & -\mathrm{id} \\ -\mathrm{id} & 0 \end{bmatrix} = \begin{bmatrix} \Phi_2^* & \bar{D}_2^* \\ \bar{m}_{T,\Phi} & \bar{m}_{T,T} \\ \bar{m}_{\Phi,\Phi} & \bar{m}_{\Phi,T} \end{bmatrix} \begin{bmatrix} \mathcal{D}_\Phi & \bar{\Delta}_\Phi \\ 0 & \mathcal{D}_D^* \end{bmatrix}\,, \tag{38}$$

for some choice of the newly introduced operators that satisfy the above identities. Note that, as in the hypotheses of Lemma 2.4, we require the operator $\bar{n}$ to be a Noether identity for the E-T-Φ_0 system that also satisfies the factorization identity given by the last equation.

Next applying Lemma 2.4 we lift the above equivalence first from the E-T-Φ_0 system to the block-triangular system (29),

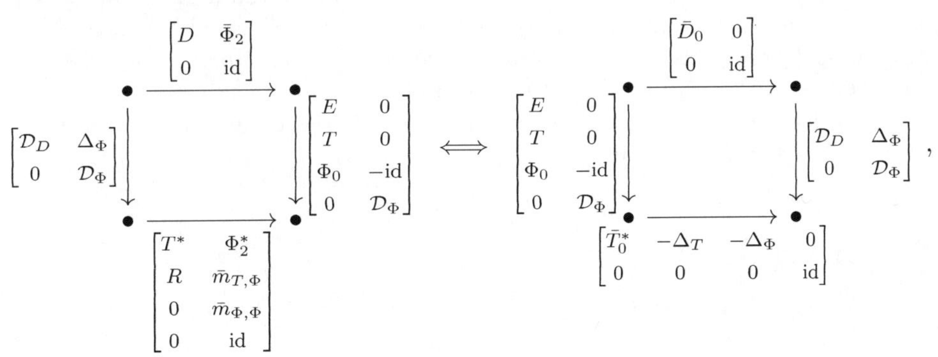

(39)

and then to the block-triangular system (32),

$$
\begin{array}{ccc}
\bullet & \xrightarrow{\begin{bmatrix} D & \bar\Phi_2 & \bar T_2 \\ 0 & \mathrm{id} & 0 \\ 0 & 0 & \mathrm{id} \end{bmatrix}} & \bullet \\
\begin{bmatrix} \mathcal{D}_D & \Delta_\Phi & \Delta_T \\ 0 & \mathcal{D}_\Phi & \bar\Delta_\Phi \\ 0 & 0 & \mathcal{D}_D^* \end{bmatrix}\Big\downarrow & & \Big\downarrow\begin{bmatrix} E & 0 & 0 \\ T & 0 & -\mathrm{id} \\ \Phi_0 & -\mathrm{id} & 0 \\ 0 & \mathcal{D}_\Phi & \bar\Delta_\Phi \\ 0 & 0 & \mathcal{D}_D^* \end{bmatrix} \\
\bullet & \xrightarrow[\begin{bmatrix} T^* & \Phi_2^* & \bar D_2^* \\ R & \bar m_{T,\Phi} & \bar m_{T,T} \\ 0 & \bar m_{\Phi,\Phi} & \bar m_{\Phi,T} \\ 0 & \mathrm{id} & 0 \\ 0 & 0 & \mathrm{id} \end{bmatrix}]{} & \bullet
\end{array}
\iff
\begin{array}{ccc}
\bullet & \xrightarrow{\begin{bmatrix} \bar D_0 & 0 & 0 \\ 0 & \mathrm{id} & 0 \\ 0 & 0 & \mathrm{id} \end{bmatrix}} & \bullet \\
\begin{bmatrix} E & 0 & 0 \\ T & 0 & -\mathrm{id} \\ \Phi_0 & -\mathrm{id} & 0 \\ 0 & \mathcal{D}_\Phi & \bar\Delta_\Phi \\ 0 & 0 & \mathcal{D}_D^* \end{bmatrix}\Big\downarrow & & \Big\downarrow\begin{bmatrix} \mathcal{D}_D & \Delta_\Phi & \Delta_T \\ 0 & \mathcal{D}_\Phi & \bar\Delta_\Phi \\ 0 & 0 & \mathcal{D}_D^* \end{bmatrix} \\
\bullet & \xrightarrow[\begin{bmatrix} \bar T_0^* & -\Delta_T & -\Delta_\Phi & 0 & 0 \\ 0 & 0 & 0 & \mathrm{id} & 0 \\ 0 & 0 & 0 & 0 & \mathrm{id} \end{bmatrix}]{} & \bullet
\end{array}
. \qquad (40)
$$

Next, to arrive at an equivalence up to homotopy with the original E equation, we repeatedly apply Lemma 2.4 and compose the appropriate morphisms from (32) and (29) to first arrive at

$$
\begin{array}{ccc}
\bullet & \xrightarrow{\begin{bmatrix} D & \bar\Phi_2 & \bar T_2 \\ 0 & 0 & \mathrm{id} \end{bmatrix}} & \bullet \\
\begin{bmatrix} \mathcal{D}_D & \Delta_\Phi & \Delta_T \\ 0 & \mathcal{D}_\Phi & \bar\Delta_\Phi \\ 0 & 0 & \mathcal{D}_D^* \end{bmatrix}\Big\downarrow & & \Big\downarrow\begin{bmatrix} E & 0 \\ T & -\mathrm{id} \\ 0 & \mathcal{D}_D^* \end{bmatrix} \\
\bullet & \xrightarrow[\begin{bmatrix} T^* & \Phi_2^* & \bar D_2^* \\ R & \bar m_{T,\Phi} & \bar m_{T,T} \\ 0 & 0 & \mathrm{id} \end{bmatrix}]{} & \bullet
\end{array}
\iff
\begin{array}{ccc}
\bullet & \xrightarrow{\begin{bmatrix} \bar D_0 & 0 \\ \Phi_0 & 0 \\ 0 & \mathrm{id} \end{bmatrix}} & \bullet \\
\begin{bmatrix} E & 0 \\ T & -\mathrm{id} \\ 0 & \mathcal{D}_D^* \end{bmatrix}\Big\downarrow & & \Big\downarrow\begin{bmatrix} \mathcal{D}_D & \Delta_\Phi & \Delta_T \\ 0 & \mathcal{D}_\Phi & \bar\Delta_\Phi \\ 0 & 0 & \mathcal{D}_D^* \end{bmatrix} \\
\bullet & \xrightarrow[\begin{bmatrix} \bar T_0^* & -\Delta_T & 0 \\ \bar\Phi_0^* & -\bar\Delta_\Phi & 0 \\ 0 & 0 & \mathrm{id} \end{bmatrix}]{} & \bullet
\end{array}
, \qquad (41)
$$

and then finally arrive at

$$
\begin{array}{ccc}
\bullet & \xrightarrow{\begin{bmatrix} \bar D_0 \\ \Phi_0 \\ T \end{bmatrix}} & \bullet \\
E\Big\downarrow & & \Big\downarrow\begin{bmatrix} \mathcal{D}_D & \Delta_\Phi & \Delta_T \\ 0 & \mathcal{D}_\Phi & \bar\Delta_\Phi \\ 0 & 0 & \mathcal{D}_D^* \end{bmatrix} \\
\bullet & \xrightarrow[\begin{bmatrix} \bar T_0^* \\ \bar\Phi_0^* \\ D^* \end{bmatrix}]{} & \bullet
\end{array}
\iff
\begin{array}{ccc}
\bullet & \xrightarrow{\begin{bmatrix} D & \bar\Phi_2 & \bar T_2 \end{bmatrix}} & \bullet \\
\begin{bmatrix} \mathcal{D}_D & \Delta_\Phi & \Delta_T \\ 0 & \mathcal{D}_\Phi & \bar\Delta_\Phi \\ 0 & 0 & \mathcal{D}_D^* \end{bmatrix}\Big\downarrow & & \Big\downarrow E \\
\bullet & \xrightarrow[\begin{bmatrix} T^* & \Phi_2^* & \bar D_2^* \end{bmatrix}]{} & \bullet
\end{array}
. \qquad (42)
$$

By construction the above morphisms are mutually on-shell inverse and in fact are an equivalence up to homotopy. The last claim would follow from Lemma 2.2(d) when neither E nor the triangular decoupled system is over-determined (Definition 2.4). In fact, in cases that are of interest to us, both systems will actually be determined.

Next, using the idea from Section 2.3, we can reduce the triangular form (42) further provided we can find δ and ε operators satisfying the identities

$$\mathcal{D}_D \circ \delta_\Phi = \Delta_\Phi + \varepsilon_\Phi \circ \mathcal{D}_\Phi \,, \qquad (43)$$

$$\mathcal{D}_\Phi \circ \bar\delta_\Phi = \bar\Delta_\Phi + \bar\varepsilon_\Phi \circ \mathcal{D}_D^* \,, \qquad (44)$$

$$\mathcal{D}_D \circ \delta_T = (\bar\Delta_T - \Delta) + \varepsilon_T \circ \mathcal{D}_D^* \,, \quad \text{where} \quad \bar\Delta_T = \Delta_T + \varepsilon_\Phi \circ \bar\Delta_\Phi \,, \qquad (45)$$

IOP Conf. Series: Journal of Physics: Conf. Series **968** (2018) 012006 doi:10.1088/1742-6596/968/1/012006

for some operator Δ. The result is the following simplified triangular form:

$$\begin{CD}
\bullet @>{\begin{bmatrix}\mathrm{id} & \delta_\Phi & \delta_T \\ 0 & \mathrm{id} & \bar{\delta}_\Phi \\ 0 & 0 & \mathrm{id}\end{bmatrix}}>> \bullet \\
@V{\begin{bmatrix}\mathcal{D}_D & \Delta_\Phi & \Delta_T \\ 0 & \mathcal{D}_\Phi & \bar{\Delta}_\Phi \\ 0 & 0 & \mathcal{D}_D^*\end{bmatrix}}VV @VV{\begin{bmatrix}\mathcal{D}_D & 0 & \Delta \\ 0 & \mathcal{D}_\Phi & 0 \\ 0 & 0 & \mathcal{D}_D^*\end{bmatrix}}V \\
\bullet @>>{\begin{bmatrix}\mathrm{id} & \varepsilon_\Phi & \varepsilon_T \\ 0 & \mathrm{id} & \bar{\varepsilon}_\Phi \\ 0 & 0 & \mathrm{id}\end{bmatrix}}> \bullet
\end{CD}$$

$$\iff \begin{CD}
\bullet @>{\begin{bmatrix}\mathrm{id} & -\delta_\Phi & -\delta_T+\delta_\Phi\circ\bar{\delta}_\Phi \\ 0 & \mathrm{id} & -\bar{\delta}_\Phi \\ 0 & 0 & \mathrm{id}\end{bmatrix}}>> \bullet \\
@V{\begin{bmatrix}\mathcal{D}_D & 0 & \Delta \\ 0 & \mathcal{D}_\Phi & 0 \\ 0 & 0 & \mathcal{D}_D^*\end{bmatrix}}VV @VV{\begin{bmatrix}\mathcal{D}_D & \Delta_\Phi & \Delta_T \\ 0 & \mathcal{D}_\Phi & \bar{\Delta}_\Phi \\ 0 & 0 & \mathcal{D}_D^*\end{bmatrix}}V \\
\bullet @>>{\begin{bmatrix}\mathrm{id} & -\varepsilon_\Phi & -\varepsilon_T+\varepsilon_\Phi\circ\bar{\varepsilon}_\Phi \\ 0 & \mathrm{id} & -\bar{\varepsilon}_\Phi \\ 0 & 0 & \mathrm{id}\end{bmatrix}}> \bullet
\end{CD} \tag{46}$$

Thus, we arrive at the final form of the equivalence up to homotopy of our original E equation with the simplified triangular form

$$\begin{CD}
\bullet @>{\begin{bmatrix}\bar{D} \\ \Phi \\ T\end{bmatrix}}>> \bullet \\
@V{E}VV @VV{\begin{bmatrix}\mathcal{D}_D & 0 & \Delta \\ 0 & \mathcal{D}_\Phi & 0 \\ 0 & 0 & \mathcal{D}_D^*\end{bmatrix}}V \\
\bullet @>>{\begin{bmatrix}\bar{T}_1^* \\ \bar{\Phi}_1^* \\ D^*\end{bmatrix}}> \bullet
\end{CD}
\iff
\begin{CD}
\bullet @>{\begin{bmatrix}D & \bar{\Phi} & \bar{T}\end{bmatrix}}>> \bullet \\
@V{\begin{bmatrix}\mathcal{D}_D & 0 & \Delta \\ 0 & \mathcal{D}_\Phi & 0 \\ 0 & 0 & \mathcal{D}_D^*\end{bmatrix}}VV @VV{E}V \\
\bullet @>>{\begin{bmatrix}T^* & \Phi_1^* & \bar{D}_1^*\end{bmatrix}}> \bullet
\end{CD}\ , \tag{47}$$

where

$$\begin{bmatrix}\bar{D} \\ \Phi \\ T\end{bmatrix} = \begin{bmatrix}\mathrm{id} & \delta_\Phi & \delta_T \\ 0 & \mathrm{id} & \bar{\delta}_\Phi \\ 0 & 0 & \mathrm{id}\end{bmatrix} \circ \begin{bmatrix}\bar{D}_0 \\ \Phi_0 \\ T\end{bmatrix}, \tag{48}$$

$$\begin{bmatrix}\bar{T}_1 \\ \bar{\Phi}_1^* \\ D^*\end{bmatrix} = \begin{bmatrix}\mathrm{id} & \varepsilon_\Phi & \varepsilon_T \\ 0 & \mathrm{id} & \bar{\varepsilon}_\Phi \\ 0 & 0 & \mathrm{id}\end{bmatrix} \circ \begin{bmatrix}\bar{T}_0^* \\ \bar{\Phi}_0^* \\ D^*\end{bmatrix}, \tag{49}$$

$$\begin{bmatrix}D & \bar{\Phi} & \bar{T}\end{bmatrix} = \begin{bmatrix}D & \bar{\Phi}_2 & \bar{T}_2\end{bmatrix} \circ \begin{bmatrix}\mathrm{id} & -\delta_\Phi & -\delta_T + \delta_\Phi \circ \bar{\delta}_\Phi \\ 0 & \mathrm{id} & -\bar{\delta}_\Phi \\ 0 & 0 & \mathrm{id}\end{bmatrix}, \tag{50}$$

IOP Conf. Series: Journal of Physics: Conf. Series **968** (2018) 012006 doi:10.1088/1742-6596/968/1/012006

$$\begin{bmatrix} T^* & \Phi_1^* & \bar{D}_1^* \end{bmatrix} = \begin{bmatrix} T^* & \Phi_2^* & \bar{D}_2^* \end{bmatrix} \circ \begin{bmatrix} \mathrm{id} & -\varepsilon_\Phi & -\varepsilon_T + \varepsilon_\Phi \circ \bar{\varepsilon}_\Phi \\ 0 & \mathrm{id} & -\bar{\varepsilon}_\Phi \\ 0 & 0 & \mathrm{id} \end{bmatrix}, \tag{51}$$

In the next section, we apply the above strategy to an explicit example.

3. Vector wave equation on Schwarzschild

Consider the exterior Schwarzschild spacetime $(\mathcal{M}, {}^4g)$ of mass $M > 0$, where $\mathcal{M} \cong \mathbb{R}^2 \times S_2$. If (S^2, Ω) is the unit round sphere and (t, r), with $-\infty < t < \infty$ and $2M < r < \infty$, are coordinates on the $\mathbb{R}^2$ factor, then the Schwarzschild metric is

$${}^4g := g + r^2\Omega\,, \quad g := -f(r)\,\mathrm{d}t^2 + \frac{\mathrm{d}r^2}{f(r)}\,, \quad f := 1 - \frac{2M}{r}\,. \tag{52}$$

For convenience we also define

$$f_1 := rf' = \frac{2M}{r}\,. \tag{53}$$

Following the 2 + 2 formalism of [11], spacetime tensors decompose into spherical and radio-temporal sectors according to

$$v_\mu \to \begin{pmatrix} v_a \\ r u_A \end{pmatrix}, \quad v^\nu \mapsto \begin{pmatrix} v^b \\ \frac{1}{r} u^B \end{pmatrix}, \tag{54}$$

which is consistent both with raising-lowering $v_\mu \leftrightarrow v^\mu$ with ${}^4g_{\mu\nu}$, as well as raising-lowering $v_a \leftrightarrow v^a$, $u_A \leftrightarrow u^A$ with g_{ab}, Ω_{AB}, respectively. If D_A, ϵ_{AB} and Y^{lm} are respectively the Levi-Civita connection, the volume form and the unit-normalized spherical harmonics on (S^2, Ω), the following tensors define a basis for the even (Y) and odd (X) vector harmonics:

$$Y_A = D_A Y\,, \quad X_A = \epsilon_{BA} D^B Y\,. \tag{55}$$

The tensor spherical harmonic decomposition of a covariant vector field is then

$$v_\mu = v_\mu^{\text{even}} + v_\mu^{\text{odd}}\,, \tag{56}$$

$$v_\mu^{\text{even}} \to \sum_{lm} \begin{pmatrix} v_a^{lm} Y^{lm} \\ r u^{lm} Y_A^{lm} \end{pmatrix}, \tag{57}$$

$$v_\mu^{\text{odd}} \to \sum_{lm} \begin{pmatrix} 0 \\ r w^{lm} X_A^{lm} \end{pmatrix}. \tag{58}$$

For conciseness, we can omit the lm indices from now on.

If ${}^4\nabla_\mu$ is the spacetime Levi-Civita connection on $(\mathcal{M}, {}^4g)$, the vector wave equation is

$${}^4\Box v_\mu := {}^4\nabla_\nu {}^4\nabla^\nu v_\mu = 0\,. \tag{59}$$

A separated mode

$$v_\mu \to \begin{pmatrix} v_t(\mathrm{d}t)_a Y + v_r(\mathrm{d}r)_a Y \\ u\, r Y_A \end{pmatrix} e^{-i\omega t} + \begin{pmatrix} 0 \\ w\, r X_A \end{pmatrix} e^{-i\omega t}, \tag{60}$$

where $v_t = v_t(r)$, $v_r = v_r(r)$, $u = u(r)$ and $w = w(r)$, satisfies the vector wave equation when

$$\boxed{1}_e \begin{bmatrix} v_t \\ v_r \\ u \end{bmatrix} := \begin{bmatrix} -\partial_r \frac{1}{f} r^2 f \partial_r v_t \\ \partial_r f r^2 f \partial_r v_r \\ \partial_r \mathcal{B}_l r^2 f \partial_r u \end{bmatrix} + \left(\frac{\omega^2}{f} - \frac{\mathcal{B}_l}{r^2}\right) \begin{bmatrix} -\frac{1}{f} r^2 v_t \\ f r^2 v_r \\ \mathcal{B}_l r^2 u \end{bmatrix}$$
$$+ i\omega r \frac{f_1}{f} \begin{bmatrix} 0 & 1 & 0 \\ -1 & 0 & 0 \\ 0 & 0 & 0 \end{bmatrix} \begin{bmatrix} v_t \\ v_r \\ u \end{bmatrix} + \begin{bmatrix} 0 & 0 & 0 \\ 0 & -2f^2 & 2\mathcal{B}_l f \\ 0 & 2\mathcal{B}_l f & \mathcal{B}_l f_1 \end{bmatrix} \begin{bmatrix} v_t \\ v_r \\ u \end{bmatrix} = 0\,, \quad (61)$$

$$\boxed{1}_o w := \partial_r \mathcal{B}_l r^2 f \partial_r w + \left(\frac{\omega^2}{f} - \frac{\mathcal{B}_l}{r^2}\right) \mathcal{B}_l r^2 w + \mathcal{B}_l f_1 w = 0\,, \quad (62)$$

where for convenience we have defined

$$\mathcal{B}_l = l(l+1)\,. \quad (63)$$

These are our *radial mode equations*. Note that we have written these equations in manifestly self-adjoint form, $\boxed{1}_e^* = \boxed{1}_e$ and $\boxed{1}_o^* = \boxed{1}_e$ (cf. Section 2.4).

We may think of the vector wave equation as the harmonic gauge-fixed Maxwell equations

$${}^4\Box v_\mu = {}^4\nabla^\nu \left({}^4\nabla_\nu v_\mu - {}^4\nabla_\mu v_\nu\right) + {}^4\nabla_\mu {}^4\nabla^\nu v_\nu\,. \quad (64)$$

We will take the harmonic gauge condition and the gauge modes to respectively be

$$\psi_0 = r\, {}^4\nabla^\mu v_\mu = 0\,, \quad v_\mu = {}^4\nabla_\mu \frac{1}{r}\phi_0\,. \quad (65)$$

After the triangular decoupling, we will show that the vector wave equation is equivalent to a system with *spin-s Regge-Wheeler (RW)* operators $\mathcal{D}_s$ [4, Eq.(23)] on the diagonal, where

$$\mathcal{D}_s \phi := \partial_r f \partial_r \phi - \frac{1}{r^2}[\mathcal{B}_l + (1-s^2) f_1]\phi + \frac{\omega^2}{f}\phi = 0\,. \quad (66)$$

In the following calculations, we will freely permit division by ω and $\mathcal{B}_l = l(l+1)$. Hence our results will hold for the $\omega \neq 0$, $l > 0$ modes. A separate treatment would be required when either of those conditions does not hold.

3.1. Odd sector

This sector is particularly simple, since it does not contain any gauge modes and is not constrained by the harmonic gauge condition. The transformation

$$\phi_1 = -i\omega r w, \quad w = -\frac{1}{i\omega r}\phi_1, \quad (67)$$

puts $\boxed{1}_o w = 0$ in direct equivalence with a $s = 1$ RW equation:

$$\boxed{1}_o w = \mathcal{B}_l i\omega r \frac{1}{\omega^2}\mathcal{D}_1 \phi_1, \quad \frac{1}{\omega^2}\mathcal{D}_1\phi_1 = \frac{1}{\mathcal{B}_l i\omega r}\boxed{1}_o w. \quad (68)$$

In diagrammatic form, we have

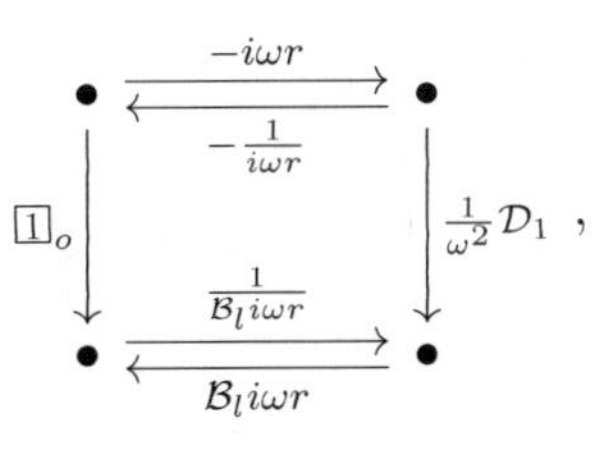

$$\left(-\frac{1}{i\omega r}\right)(-i\omega r) = \mathrm{id}\,, \qquad \left(\frac{1}{\mathcal{B}_l i\omega r}\right)(\mathcal{B}_l i\omega r) = \mathrm{id}\,. \quad (69)$$

3.2. Even sector

This sector is more complicated and is the first prototype for the abstract approach to triangular decoupling that we outlined earlier in Section 2.5. Let us now explicitly introduce some of the auxiliary operators that we will need, following the notation established earlier:

$$E := \boxed{1}_e \,,$$

$$D := \frac{1}{\omega^2 r^2}\begin{bmatrix} -i\omega r \\ r^2\partial_r \frac{1}{r} \\ 1 \end{bmatrix}, \qquad T^* := \begin{bmatrix} \frac{i\omega r}{f} \\ f r^2 \partial_r \frac{1}{r} \\ \mathcal{B}_l \end{bmatrix},$$

$$T := \begin{bmatrix} -\frac{i\omega r}{f} & -\frac{1}{r}\partial_r f r^2 & \mathcal{B}_l \end{bmatrix}, \qquad D^* := \frac{1}{\omega^2 r^2}\begin{bmatrix} i\omega r & -r\partial_r & 1 \end{bmatrix},$$

$$\mathcal{D}_D := \frac{1}{\omega^2}\mathcal{D}_0 \,, \qquad \mathcal{D}_D^* := \frac{1}{\omega^2}\mathcal{D}_0 \,,$$

$$\Phi_0 := \begin{bmatrix} 0 & -f & f\partial_r r \end{bmatrix}, \qquad \bar{\Phi}_0^* := \frac{1}{r^2}\begin{bmatrix} 0 & -1 & \frac{1}{\mathcal{B}_l} r^2 \partial_r \frac{f}{r} \end{bmatrix},$$

$$\mathcal{D}_\Phi := \frac{1}{\omega^2}\mathcal{D}_1 \,, \qquad R = \bar{R} := -\mathrm{id} \,.$$

They can be obtained from a mode decomposition of the spacetime operators we discussed earlier, up to simple multiplicative normalizations. We have chosen the normalizations so that all operators have dimensionless components, under the convention that each of r^{-1}, ω, and ∂_r has the dimension of inverse length, while f, f_1 and $\mathcal{B}_l$ are dimensionless. The operators

$$\begin{bmatrix} v_t \\ v_r \\ u \end{bmatrix} = D[\phi_0], \quad \phi_1 = \Phi_0 \begin{bmatrix} v_t \\ v_r \\ u \end{bmatrix}, \quad \psi_0 = T \begin{bmatrix} v_t \\ v_r \\ u \end{bmatrix} \tag{70}$$

respectively separate out the *gauge modes* (ϕ_0), the *gauge invariant modes* (ϕ_1) and the *constraint violating modes* (ψ_0).

Computing the composition

$$\bar{\Delta}_\Phi := \bar{\Phi}_0^* \circ T^* \circ \bar{R} = -\frac{f_1}{\omega^2 r^2}$$

is enough to get us to the E-T-Φ_0 triangular form (32).

Next, we must invert the morphism D of gauge modes satisfying $\mathcal{D}_D[\phi_0] = 0$ into the E-T-Φ_0 system. At this step it is helpful to use a computer algebra system. Any equivalence class of scalar operators acting on the v_t, v_r, u variables modulo the E-T-Φ_0 equations can be uniquely represented by a linear combination of v_t and $\partial_r v_t$ (all second and higher derivatives are eliminated by E, $\partial_r v_r$ is eliminated by T and $\partial_r u$ is eliminated by Φ_0). Precomposing each of them with D, we get representatives of equivalence classes of scalar operators acting on the ϕ_0 variable modulo the $\mathcal{D}_D$ equation. These, in turn, are uniquely represented by ϕ_0 and $\partial_r\phi_0$. Algebraically inverting the relationship between $v_t, \partial_r v_t$ and $\phi_0, \partial_r\phi_0$ gives us the operator $\bar{D}_0$ from (32). Keeping track of how higher derivatives are eliminated by the equations also allows us to complete $\bar{D}_0$ to a cochain map. The result is

$$\bar{D}_0 = \begin{bmatrix} i\omega r & 0 & 0 \end{bmatrix},$$

$$\begin{bmatrix} \bar{T}_0^* & -\Delta_T & -\Delta_\Phi \end{bmatrix} = \frac{1}{\omega^2 r^2}\begin{bmatrix} -i\omega r f & -f_1 & 0 \mid -f_1 f r^2 \partial_r \frac{1}{r} \mid \mathcal{B}_l f_1 \end{bmatrix}.$$

IOP Conf. Series: Journal of Physics: Conf. Series **968** (2018) 012006 doi:10.1088/1742-6596/968/1/012006

Further explicit calculations give us the homotopy corrections needed to proceed with the decoupling:

$$h_D = 0\,, \tag{71}$$

$$\begin{bmatrix} \bar{h}_E & \bar{T}_2 & \bar{\Phi}_2 \end{bmatrix} = \frac{1}{\omega^2 r^2} \left[\begin{array}{ccc|c|c} 0 & 0 & 0 & 0 & 0 \\ 0 & 1 & 0 & f r^2 \partial_r \frac{1}{r} & -\mathcal{B}_l \\ 0 & 0 & \frac{f}{\mathcal{B}_l} & f & -f r \partial_r \end{array}\right], \tag{72}$$

$$\bar{n} = \left[\begin{array}{cc} 0 & 0 \\ -f r^2 \partial_r \frac{f}{r} & \mathcal{B}_l f \\ -\mathcal{B}_l f & \mathcal{B}_l r \partial_r f \\ \hline f & 0 \\ \hline 0 & f \end{array}\right] \begin{bmatrix} -D^* & \mathcal{D}_D^* & 0 \\ -\bar{\Phi}_0^* & \bar{\Delta}_\Phi & \mathcal{D}_\Phi \end{bmatrix}, \tag{73}$$

$$\begin{bmatrix} \Phi_2^* & \bar{D}_2^* \\ \bar{m}_{T,\Phi} & \bar{m}_{T,T} \\ \bar{m}_{\Phi,\Phi} & \bar{m}_{\Phi,T} \end{bmatrix} = \left[\begin{array}{cc} 0 & 0 \\ -\mathcal{B}_l f & f r^2 \partial_r \frac{f}{r} \\ -\mathcal{B}_l r \partial_r f & \mathcal{B}_l f \\ \hline 0 & -f \\ \hline -f & 0 \end{array}\right]. \tag{74}$$

From the above operators, we can construct the desired triangular decoupling (42) of E:

$$\begin{bmatrix} \bar{D}_0 \\ \Phi_0 \\ T \end{bmatrix} = \left[\begin{array}{ccc} i\omega r & 0 & 0 \\ \hline 0 & -f & f \partial_r r \\ \hline -\frac{i\omega r}{f} & -\frac{1}{r} \partial_r f r^2 & \mathcal{B}_l \end{array}\right], \tag{75}$$

$$\begin{bmatrix} \bar{T}_0^* \\ \bar{\Phi}_0^* \\ D^* \end{bmatrix} = \frac{1}{\omega^2 r^2} \left[\begin{array}{ccc} -i\omega r f & -f_1 & 0 \\ \hline 0 & -1 & \frac{1}{\mathcal{B}_l} r^2 \partial_r \frac{f}{r} \\ \hline i\omega r & -r\partial_r & 1 \end{array}\right], \tag{76}$$

$$\begin{bmatrix} D & \bar{\Phi}_2 & \bar{T}_2 \end{bmatrix} = \frac{1}{\omega^2 r^2} \left[\begin{array}{c|c|c} -i\omega r & 0 & 0 \\ r^2 \partial_r \frac{1}{r} & -\mathcal{B}_l & f r^2 \partial_r \frac{1}{r} \\ 1 & -f r \partial_r & f \end{array}\right], \tag{77}$$

$$\begin{bmatrix} T^* & \Phi_2^* & \bar{D}_2^* \end{bmatrix} = \left[\begin{array}{c|c|c} \frac{i\omega r}{f} & 0 & 0 \\ f r^2 \partial_r \frac{1}{r} & -\mathcal{B}_l f & f r^2 \partial_r \frac{f}{r} \\ \mathcal{B}_l & -\mathcal{B}_l r \partial_r f & \mathcal{B}_l f \end{array}\right]. \tag{78}$$

An explicit calculation also shows that the cochain maps that we have obtained are mutually inverse up to homotopy:

$$\begin{bmatrix} \bar{D}_0 \\ \Phi_0 \\ T \end{bmatrix} \circ \begin{bmatrix} D & \bar{\Phi}_2 & \bar{T}_2 \end{bmatrix} = \mathrm{id} - \begin{bmatrix} 0 & 0 & 0 \\ 0 & f & 0 \\ 1 & 0 & f \end{bmatrix} \circ \begin{bmatrix} \mathcal{D}_D & \Delta_\Phi & \Delta_T \\ 0 & \mathcal{D}_\Phi & \bar{\Delta}_\Phi \\ 0 & 0 & \mathcal{D}_D^* \end{bmatrix}, \tag{79}$$

$$\begin{bmatrix} \bar{D}_0 \\ \Phi_0 \\ T \end{bmatrix} \circ \begin{bmatrix} D & \bar{\Phi}_2 & \bar{T}_2 \end{bmatrix} = \mathrm{id} - \frac{1}{\omega^2 r^2} \begin{bmatrix} 0 & 0 & 0 \\ 0 & 1 & 0 \\ 0 & 0 & \frac{f}{\mathcal{B}_l} \end{bmatrix} \circ \square_e\,. \tag{80}$$

IOP Conf. Series: Journal of Physics: Conf. Series **968** (2018) 012006 doi:10.1088/1742-6596/968/1/012006

Noting the helpful identities

$$\frac{1}{\omega^2}\mathcal{D}_{s_1}\frac{1}{(s_1^2-s_2^2)} = \frac{f_1}{r^2} + \frac{1}{(s_1^2-s_2^2)}\frac{1}{\omega^2}\mathcal{D}_{s_2}\,, \tag{81}$$

$$\frac{1}{\omega^2}\mathcal{D}_0\left(\frac{f}{2}\right) = \left(\frac{f_1}{\omega^2 r^2} f r^2 \partial_r \frac{1}{r} + \frac{f_1^2}{2\omega^2 r^2}\right) + \left(\frac{f}{2}\right)\frac{1}{\omega^2}\mathcal{D}_0\,, \tag{82}$$

we can use the idea of Section 2.3 to further reduce the triangular form with the following operators:

$$\bar\delta_\Phi = -\mathrm{id}\,, \qquad \delta_\Phi = \mathcal{B}_l\,, \qquad \delta_T = \frac{f}{2}\,,$$
$$\bar\varepsilon_\Phi = -\mathrm{id}\,, \qquad \varepsilon_\Phi = \mathcal{B}_l\,, \qquad \varepsilon_T = \frac{f}{2}\,,$$

which are needed to kill or simplify the off-diagonal terms

$$\bar\Delta_\Phi = -\frac{f_1}{\omega^2 r^2}\,, \tag{83}$$

$$\Delta_\Phi = -\frac{f_1}{\omega^2 r^2}\mathcal{B}_l\,, \tag{84}$$

$$\Delta_T = \frac{f_1}{\omega^2 r^2} f r^2 \partial_r \frac{1}{r}\,, \tag{85}$$

$$\text{and}\quad \bar\Delta_T = \Delta_T + \varepsilon_\Phi \circ \bar\Delta_\Phi = \frac{f_1}{\omega^2 r^2}\left(f r^2 \partial_r \frac{1}{r} - \mathcal{B}_l\right). \tag{86}$$

The remaining off-diagonal term in the triangular form (46) is then

$$\Delta = \bar\Delta_T - \left(\frac{f_1}{\omega^2 r^2} f r^2 \partial_r \frac{1}{r} + \frac{f_1^2}{2\omega^2 r^2}\right) = -\frac{f_1}{\omega^2 r^2}\left(\mathcal{B}_l + \frac{f_1}{2}\right), \tag{87}$$

so that our final reduced triangular form is

$$\boxed{1}_e \begin{bmatrix} v_t \\ v_r \\ u \end{bmatrix} = 0 \iff \frac{1}{\omega^2}\begin{bmatrix} \mathcal{D}_0 & 0 & -\frac{f_1}{r^2}\left(\mathcal{B}_l + \frac{f_1}{2}\right) \\ 0 & \mathcal{D}_1 & 0 \\ 0 & 0 & \mathcal{D}_0 \end{bmatrix}\begin{bmatrix} \phi_0 \\ \phi_1 \\ \psi_0 \end{bmatrix} = 0\,. \tag{88}$$

The following operators will provide us with the final pieces necessary to construct a homotopy equivalence between $\boxed{1}_e$ and the reduced triangular form (88):

$$\begin{bmatrix} \phi_0 \\ \phi_1 \\ \psi_0 \end{bmatrix} = \begin{bmatrix} \bar D \\ \Phi \\ T \end{bmatrix}\begin{bmatrix} v_t \\ v_r \\ u \end{bmatrix} = \begin{bmatrix} 1 & \mathcal{B}_l & \frac{f}{2} \\ 0 & 1 & -1 \\ 0 & 0 & 1 \end{bmatrix} \circ \left[\begin{array}{ccc} i\omega r & 0 & 0 \\ \hline 0 & -f & f\partial_r r \\ \hline -\frac{i\omega r}{f} & -\frac{1}{r}\partial_r f r^2 & \mathcal{B}_l \end{array}\right]\begin{bmatrix} v_t \\ v_r \\ u \end{bmatrix}, \tag{89}$$

$$\begin{bmatrix} \bar T_1 \\ \bar\Phi_1^* \\ D^* \end{bmatrix} = \begin{bmatrix} 1 & \mathcal{B}_l & \frac{f}{2} \\ 0 & 1 & -1 \\ 0 & 0 & 1 \end{bmatrix} \circ \frac{1}{\omega^2 r^2}\left[\begin{array}{ccc} -i\omega r f & -f_1 & 0 \\ \hline 0 & -1 & \frac{1}{\mathcal{B}_l} r^2 \partial_r \frac{f}{r} \\ \hline i\omega r & -r\partial_r & 1 \end{array}\right], \tag{90}$$

$$\begin{bmatrix} v_t \\ v_r \\ u \end{bmatrix} = \begin{bmatrix} D & \bar\Phi & \bar T \end{bmatrix}\begin{bmatrix} \phi_0 \\ \phi_1 \\ \psi_0 \end{bmatrix} = \frac{1}{\omega^2 r^2}\left[\begin{array}{c|c|c} -i\omega r & 0 & 0 \\ r^2\partial_r \frac{1}{r} & -\mathcal{B}_l & f r^2 \partial_r \frac{1}{r} \\ 1 & -f r \partial_r & f \end{array}\right] \circ \begin{bmatrix} 1 & -\mathcal{B}_l & -\frac{f}{2} - \mathcal{B}_l \\ 0 & 1 & 1 \\ 0 & 0 & 1 \end{bmatrix}\begin{bmatrix} \phi_0 \\ \phi_1 \\ \psi_0 \end{bmatrix}, \tag{91}$$

$$\begin{bmatrix} T^* & \Phi_1^* & \bar D_1^* \end{bmatrix} = \left[\begin{array}{c|c|c} \frac{i\omega r}{f} & 0 & 0 \\ f r^2 \partial_r \frac{1}{r} & -\mathcal{B}_l f & f r^2 \partial_r \frac{f}{r} \\ \mathcal{B}_l & -\mathcal{B}_l r \partial_r f & \mathcal{B}_l f \end{array}\right] \circ \begin{bmatrix} 1 & -\mathcal{B}_l & -\frac{f}{2} - \mathcal{B}_l \\ 0 & 1 & 1 \\ 0 & 0 & 1 \end{bmatrix}. \tag{92}$$

IOP Conf. Series: Journal of Physics: Conf. Series **968** (2018) 012006 doi:10.1088/1742-6596/968/1/012006

4. Discussion

We have shown how an abstract strategy, based on formal properties of linear differential operators, can be used to decouple the radial mode equations (with nonzero frequency ω and angular momentum l) of the vector wave equation on Schwarzschild spacetime into a system of hierarchically (that is *triangularly*) coupled scalar equations. Essentially, we have explicitly separated out the degrees of freedom of the vector field into *pure gauge*, *gauge invariant* and *constraint violating* modes (with the interpretation that the vector wave equation arose as the harmonic (or Lorenz) gauge fixed Maxwell equations). This strategy was first successfully applied in [7]. However, it should be noted that this decoupling strategy is never made explicit in [7], where it becomes apparent only as a common pattern in the voluminous explicit calculations in a sequence of examples of increasing complexity. One of the main contributions of our work is to make this decoupling strategy explicit and mathematically founded on basic ideas drawn from the theory of D-modules [8, Sec.10.5], the category theoretic approach to differential equations [9, Sec.VII.5] and homological algebra [10].

It is worth pointing out that, even though our approach to triangular decoupling and reduction of the triangular form owes much to [7], the triangular form (88) appears in [7] only implicitly. Moreover, the reduction of the triangular form in [7] only kills the Δ_Φ (84) off-diagonal term. Since we have also killed the $\bar{\Delta}_\Phi$ (83) and simplified the Δ_T (85) terms, our decoupling goes a step further.

The next step is to apply the same decoupling strategy to linearized gravity in harmonic (or de Donder) gauge on the Schwarzschild spacetime. In fact, this strategy was already successfully applied to that case in [7], which was actually one of the main goals of that work. Unfortunately, this decoupling for linearized gravity was achieved only after some rather voluminous explicit calculations and the final results were not presented in a very economic way. We believe that following the abstract strategy in Section 2.5, can significantly streamline both the presentation of the final results and the intermediate calculations. The details will be discussed elsewhere.

We believe that the same strategy, of triangular decoupling by separating out *purge gauge*, *gauge invariant* and *constraint violating* modes should also work for harmonic gauge field equations (either Maxwell or linearized gravity) on other spacetime geometries where the physical degrees of freedom satisfy decoupled equations. The Kerr spacetime is an example, where at the mode level the gauge invariant degrees of freedom decouple and satisfy Teukolsky or Fackerell-Ipser equations [1, 2]. The main uncertainty there is the degree to which the triangular form can be explicitly reduced using the method of Section 2.3.

As was explained in the Introduction, the complicated form of the radial mode equations (61) and (62) of the vector wave equation is a significant obstacle to the study of their analytical properties, and in particular their spectral theory. Once these equations have been put into the decoupled triangular form (88) and (68), their analytical properties can be more easily deduced from their diagonal parts. These diagonal parts consist of spin-s Regge-Wheeler equations, whose analytical properties are well understood [1]. Such information can then be used to construct both classical and quantum propagators (or Green functions) for the vector wave equation on Schwarzschild spacetime. The same will be true for linearized gravity, once a similar decoupling has been achieved. For that purpose, the $\omega = 0$ and $l = 0$ modes would need to be investigated as well. These applications will also be discussed elsewhere.

Finally, it would also be interesting to relate our decoupling strategy to the *adjoint operator method* of Wald [12, 13]. Recall that we have liberally used formal adjoints (Section 2.4) as part of our notation. This was convenient because some of the operators that we used in the case of the vector wave equation, in Section 3, really were formal adjoints of each other. However, we did not really use this property as part of our decoupling strategy. On the other hand, supposing that we can find a decoupling of a formally self-adjoint differential equation $e = e^*$ into a triangular form $\bar{e}$, taking adjoints produces for free another decoupling of e but this time into $\bar{e}^*$. This is illustrated in the following commutative diagrams:

$$\begin{array}{ccc} \bullet & \underset{\bar{k}}{\overset{k}{\rightleftarrows}} & \bullet \\ {\scriptstyle e}\downarrow & & \downarrow{\scriptstyle \bar{e}} \\ \bullet & \underset{\bar{g}}{\overset{g}{\rightleftarrows}} & \bullet \end{array} \quad \Longleftrightarrow \quad \begin{array}{ccc} \bullet & \underset{g^*}{\overset{\bar{g}^*}{\rightleftarrows}} & \bullet \\ {\scriptstyle e=e^*}\downarrow & & \downarrow{\scriptstyle \bar{e}^*} \\ \bullet & \underset{k^*}{\overset{\bar{k}^*}{\rightleftarrows}} & \bullet \end{array} . \qquad (93)$$

If $\bar{e}$ is in upper triangular form, then $\bar{e}^*$ is in lower triangular form. In the vector wave equation example (88), the adjoint equation $\bar{e}^*$ is actually equivalent to the original equation $\bar{e}$ by the simple interchange $\phi_0 \leftrightarrow \psi_0$. In such a case, composing the equivalences of e with $\bar{e}$, of $\bar{e}$ with $\bar{e}^*$, and of $\bar{e}^*$ with $e^* = e$, we obtain a morphism from e to itself, or more precisely a pair of operators $\tilde{k}, \tilde{g}$ satisfying $e \circ \tilde{k} = \tilde{g} \circ e$. In the terminology of [13], $\tilde{k}$ is a *symmetry operator* and it maps solutions of e to solutions of e (in this case, isomorphically). Thus, if $\tilde{k}$ is not simply proportional to a constant, it gives an interesting way to generate new solutions from known ones. The significance of the existence of such a symmetry operator should be investigated further.

Acknowledgments

The author was in part supported by the National Group of Mathematical Physics (GNFM-INdAM). The author also thanks Francesco Bussola for checking some of this paper's calculations as part of his MSc thesis [14].

References

[1] Chandrasekhar S 1983 *The mathematical theory of black holes* (*International series of monographs on physics* vol 69) (Oxford: Clarendon Press) ISBN 9780198512912
[2] Frolov V and Novikov I 1998 *Black Hole Physics: Basic Concepts and New Developments* (*Fundamental Theories of Physics* vol 96) (Dordrecht: Kluwer Academic Publishers) ISBN 0792351460
[3] Poisson E, Pound A and Vega I 2011 *Living Reviews in Relativity* **14** (*Preprint* 1102.0529)
[4] Rosa J G and Dolan S R 2012 *Physical Review D* **85** 044043 (*Preprint* 1110.4494)
[5] Fewster C J and Hunt D S 2013 *Reviews in Mathematical Physics* **25** 1330003 (*Preprint* 1203.0261)
[6] Benini M, Dappiaggi C and Murro S 2014 *Journal of Mathematical Physics* **55** 082301 (*Preprint* 1404.4551)
[7] Berndtson M V 2007 *Harmonic gauge perturbations of the Schwarzschild metric* Ph.D. thesis University of Colorado (*Preprint* 0904.0033)
[8] Seiler W M 2010 *Involution: The Formal Theory of Differential Equations and its Applications in Computer Algebra* (*Algorithms and Computation in Mathematics* vol 24) (Berlin: Springer)
[9] Krasilshchik I S, Lychagin V V and Vinogradov A M 1986 *Geometry of jet spaces and nonlinear partial differential equations* Advanced studies in contemporary mathematics (New York, NY: Gordon and Breach) ISBN 2881240518
[10] Weibel C A 1994 *An introduction to homological algebra* (*Cambridge Studies in Advanced Mathematics* vol 38) (Cambridge: Cambridge University Press) ISBN 9781139644136
[11] Martel K and Poisson E 2005 *Physical Review D* **71** 104003 (*Preprint* gr-qc/0502028)
[12] Wald R M 1978 *Physical Review Letters* **41** 203–206
[13] Aksteiner S and Bäckdahl T 2016 Symmetries of linearized gravity from adjoint operators (*Preprint* 1609.04584)
[14] Bussola F 2015 *De Donder gauge graviton Green's function in Schwarzschild spacetime with an outlook toward the Feynman propagator* Master's thesis University of Trento

IOP Conf. Series: Journal of Physics: Conf. Series **968** (2018) 012007 doi:10.1088/1742-6596/968/1/012007

Connes distance and optimal transport

Pierre Martinetti

Dipartimento di matematica, università di Genova, via Dodecaneso 35, 16146 Genova

E-mail: martinetti@dima.unige.it

Abstract. We give a brief overview on the relation between Connes spectral distance in noncommutative geometry and the Wasserstein distance of order 1 in optimal transport. We first recall how these two distances coincide on the space of probability measures on a Riemannian manifold. Then we work out a simple example on a discrete space, showing that the spectral distance between arbitrary states does not coincide with the Wasserstein distance with cost the spectral distance between pure states.

1. The metric aspect of noncommutative geometry

Topology is the minimal structure required for a set to be called *space*, and for its elements to become *points*. Topology gives sense to the notion of neighbourhood (and more generally to the notion of open sets). The algebraic dual notion is that of continuity (a continuous function being, by definition, such that the inverse image of any open set is an open set). Actually this duality goes quite far, since all the information of a topological space is contained within the algebra of continuous functions defined on it. More precisely, Gelfand's duality states that any complex *commutative* C^*-algebra $\mathcal{A}$ is isomorphic to the algebra of continuous functions vanishing at infinity on some locally compact topological space - the space $\mathcal{P}(\mathcal{A})$ of pure states of $\mathcal{A}$ - and conversely any locally compact topological space $\mathcal{X}$ is homeomorphic to the space of pure states of the *commutative* algebra $C_0(\mathcal{X})$ of continuous functions on $\mathcal{X}$ vanishing at infinity:

$$\mathcal{A} \simeq C_0(\mathcal{P}(\mathcal{A})), \quad \mathcal{X} \simeq \mathcal{P}(C_0(\mathcal{X})). \tag{1}$$

Recall that the pure states of an involutive algebra $\mathcal{A}$ are the extremal points of the set of states, the latter being the linear maps φ on $\mathcal{A}$ wich are positive - $\varphi(a^*a) \in \mathbb{R}^+$ - and of norm 1 (where $||\varphi|| = \sup_{a\in\mathcal{A}} \frac{|\varphi(a)|}{||a||}$). In particular, a state of the commutative algebra $C_0(\mathcal{X})$ is the integration with respect to a probability measure μ, with pure states given by Dirac δ measures (i.e. evaluation at a point):

$$\mathcal{P}(C_0(\mathcal{X})) \ni \delta_x : x \to \int_{\mathcal{X}} f\delta_x = f(x)\,, \qquad S(C_0(\mathcal{X})) \ni \varphi : f \to \int_{\mathcal{X}} f\, \mathrm{d}\mu \quad \forall f \in C_0(\mathcal{X}).$$

Connes' noncommutative geometry extends Gelfand's duality beyond topology, so that to encompass all the aspects of Riemannian geometry, in particular the metric. To do so, one needs more than an algebra: a *spectral triple* [2] consists in an involutive algebra $\mathcal{A}$ acting faithfully on an Hilbert space $\mathcal{H}$, with D a selfadjoint operator on $\mathcal{H}$ such that the commutator $[D, a]$ is bounded and $a[D - \lambda\mathbb{I}]^{-1}$ is compact for any $a \in \mathcal{A}$ and $\lambda \notin \mathrm{Sp}\, D$. When a set of conditions (dimension, regularity, finitude, first order, orientability) is satisfied, then one is able to characterize a Riemannian manifold by purely spectral data [5]:

IOP Conf. Series: Journal of Physics: Conf. Series **968** (2018) 012007 doi:10.1088/1742-6596/968/1/012007

- For $\mathcal{M}$ a compact Riemannian manifold, then

$$(C^\infty(\mathcal{M}), \Omega^\bullet(\mathcal{M}), d + d^\dagger) \tag{2}$$

is a spectral triple, where $\Omega^\bullet(\mathcal{M})$ is the Hilbert space of square integrable differential forms on $\mathcal{M}$ and $d{+}d^\dagger$ is the Hodge-Dirac operator (d the exterior derivative, $d^\dagger$ its Hodge-adjoint).
- When $(\mathcal{A}, \mathcal{H}, D)$ is a spectral triple with $\mathcal{A}$ unital commutative, then there exists a compact Riemannian manifold $\mathcal{M}$ such that $\mathcal{A} = C^\infty(\mathcal{M})$.

By adding two extra-conditions (real structure and Poincaré duality), the result is extended to spin manifolds.

Why is such a spectral characterization of manifolds interesting ? Because the properties defining a spectral triple still make sense for a noncommutative $\mathcal{A}$ [2]. A *noncommutative geometry* is thus a spectral triple where the algebra $\mathcal{A}$ is noncommutative. At the light of Gelfand duality, this is the geometrical object whose algebra of functions defined on it is non commutative. As such it cannot be a usual topological space (otherwise its algebra of continuous functions would be commutative), but it rather appears as a "space without points".

commutative spectral triple	$\rightarrow$	noncommutative spectral triple
$\updownarrow$		$\downarrow$
Riemannian geometry		non-commutative geometry

However, always at the light of Gelfand duality, it is tempting to consider the pure states of the noncommutative algebra as the equivalent of points in the noncommutative context. This is all the more appealing that the same formula that allows to retrieve the Riemannian geodesic distance in Connes reconstruction theorem, also provides the space of states with a distance.

Explicitly, given a spectral triple $(\mathcal{A}, \mathcal{H}, D)$, with $\mathcal{A}$ commutative or not, one defines on its state space $\mathcal{S}(\mathcal{A})$ the *spectral distance* [3]

$$d_D(\varphi, \psi) = \sup_{a \in \mathcal{A}} \{ |\varphi(a) - \psi(a)| \,/\, \|[D, a]\| \leq 1 \} \qquad \forall \varphi, \psi \in \mathcal{S}(\mathcal{A}). \tag{3}$$

It is not difficult to check that d_D has all the properties of a distance (zero if and only if $\varphi = \psi$, $d_D(\varphi, \psi) = d_D(\psi, \varphi)$, triangle inequality), except that it may be infinite. By a slight abuse of language, we still call it *distance*. For an overview of explicit computation of this distance in various examples of commutative and noncommutative spectral triples, see [11].

2. Rieffel's remark and Wasserstein distance of order 1

Rieffel noticed in [12] that formula (3), applied to the spectral triple of a Riemannian manifold (2), was nothing but the Wasserstein distance of order 1 in the theory of optimal transport, or more exactly a reformulation of Kantorovich dual of the Wasserstein distance.

To see it, let us first remind what the Wasserstein (or Monge Kantorovich distance) is. Let $\mathcal{X}$ be a locally compact Polish space, $c(x, y)$ a positive real function, the "cost". The minimal work W required to transport the probability measure μ_1 to μ_2 is

$$W(\mu_1, \mu_2) \doteq \inf_\pi \int_{\mathcal{X} \times \mathcal{X}} c(x, y)\, d\pi \tag{4}$$

IOP Conf. Series: Journal of Physics: Conf. Series **968** (2018) 012007 doi:10.1088/1742-6596/968/1/012007

where the infimum is over all *transportation plans*, i.e. measures π on $\mathcal{X} \times \mathcal{X}$ with marginals μ_1, μ_2. When the cost function c is a distance d, then

$$W(\mu_1, \mu_2) \doteq \inf_{\pi} \int_{\mathcal{X}\times\mathcal{X}} d(x, y)\, d\pi \tag{5}$$

is a distance (possibly infinite) on the space of probability measures on $\mathcal{X}$, called the Monge-Kantorovich (or Wasserstein) distance of order 1.

In [9], Kantorovich showed that Monge problem of minimizing the cost (eq. (4)) had an equivalent dual formulation (interpreted as maximizing a profit). Namely, $W(\mu_1, \mu_2)$ is equal to

$$W(\varphi_1, \varphi_2) = \sup_{\|f\|_{\text{Lip}} \leq 1} \left(\int_{\mathcal{X}} f d\mu_1 - \int_{\mathcal{X}} f d\mu_2 \right) \tag{6}$$

where φ_1, φ_2 are the states of $C(\mathcal{X})$ defined by the measure μ_1, μ_2:

$$\varphi_i(f) = \int_{\mathcal{X}} f d\mu_i \quad \forall f \in C(\mathcal{X}),\ i = 1, 2; \tag{7}$$

and the supremum in (6) is on all the functions 1-Lipschitz with respect to the cost, that is

$$f(x, y) \leq c(x, y) \quad \forall x, y \in \mathcal{X}. \tag{8}$$

Let $\mathcal{X} = \mathcal{M}$ be a complete, connected, without boundary, Riemannian manifold. For any $\varphi, \tilde{\varphi} \in \mathcal{S}(C_0(\mathcal{M}))$,

$$W(\varphi, \tilde{\varphi}) = d_D(\varphi, \tilde{\varphi})$$

where W is the Wasserstein distance associated to the cost d_{geo}, while d_D is the spectral distance associated to $\left(C_0^\infty(\mathcal{M}), \Omega^\bullet(\mathcal{M}), D = d + d^\dagger\right)$. That Kantorovich dual (6) coincides with the spectral distance (3) then follows from the observation that the supremum in the latter can be searched equivalently on selfadjoint elements, for which one has

$$\left\| [d + d^\dagger, f]^2 \right\| = \|f\|^2_{\text{Lip}}\,. \tag{9}$$

As pointed out in [7], one has to be careful that the manifold is complete, otherwise there is no guarantee that the supremum on the 1-Lipschitz functions in (6) coincides with the supremum on 1-Lipschitz functions vanishing at infinity in (3).

3. Towards a theory of optimal transport in noncommutative geometry ?

Connes spectral distance on a manifold coincides with Kantorovich dual formulation of the Wasserstein distance of order 1. It is quite natural to wonder if the same is true is the noncommutative setting. But there does not exist any "noncommutative Wasserstein distance" of whom the spectral distance would be the dual. Is it possible to build one ? More specifically, given a spectral triple $(\mathcal{A}, \mathcal{H}, D)$ with noncommutative $\mathcal{A}$, is there a Wasserstein distance W_D on $\mathcal{S}(\mathcal{A})$ such that its Kantorovich dual is the spectral distance d_D ?

Commutative case:		**Noncommutative case:**
Connes distance d_D	$\rightarrow$	Connes distance d_D
$\uparrow$		$\vert$
Kantorovich duality		Kantorovich duality ?
$\downarrow$		$\downarrow$
Wassertein distance W with $d_D(\delta_x, \delta_y)$ as a cost function		W_D for some noncommutative cost ?

IOP Conf. Series: Journal of Physics: Conf. Series **968** (2018) 012007 doi:10.1088/1742-6596/968/1/012007

In the commutative case $\mathcal{A} = C^\infty(\mathcal{M})$, one retrieves the cost function as the Wasserstein distance between pure states:

$$W(\delta_x, \delta_y) = c(x, y).$$

So it is tempting to define W_D on the whole space of states $\mathcal{S}(\mathcal{A})$ as the Wasserstein distance associated with the cost c defined by the spectral distance on the space of pure states $\mathcal{P}(\mathcal{A})$, that is

$$c(\omega_1, \omega_2) := d_D(\omega_1, \omega_2) \tag{10}$$

We worked out this construction in [10], restricting to unital separable C^*-algebras, for which it is known that a state is a probability measure on $\mathcal{P}(\mathcal{A})$ [1, p.144] [1], namely to any $\varphi \in \mathcal{S}(\mathcal{A})$, there exists a (non-necessarily unique) probability measure $\mu \in \text{Prob}(\mathcal{P}(\mathcal{A}))$ such that

$$\varphi(a) = \int_{\mathcal{P}(\mathcal{A})} \hat{a}(\omega)\, d\mu(\omega) \tag{11}$$

where $\hat{a}(\omega) \doteq \omega(a)$ denotes the evaluation at $\omega \in \mathcal{P}(\mathcal{A})$ of an element a of $\mathcal{A}$, viewed as a function on $\mathcal{P}(\mathcal{A})$. The Wasserstein distance on $\mathcal{S}(\mathcal{A})$ associated with the cost (10) has Kantorovich-dual formulation

$$W_D(\varphi, \tilde{\varphi}) \doteq \sup_{a \in \text{Lip}_D(\mathcal{A})} \left\{ \left| \int_{\mathcal{P}(\mathcal{A})} \hat{a}(\omega)\, d\mu(\omega) - \int_{\mathcal{P}(\mathcal{A})} \hat{a}(\omega)\, d\tilde{\mu}(\omega) \right| \right\}, \tag{12}$$

where

$$\text{Lip}_D(\mathcal{A}) \doteq \{a \in \mathcal{A} \text{ such that } |\omega_1(a) - \omega_2(a)| \leq d_D(\omega_1, \omega_2)\ \forall \omega_1, \omega_2 \in \mathcal{P}(\mathcal{A})\} \tag{13}$$

is the set of element of $\mathcal{A}$ that are 1-Lipschitz with respect to the cost (10).

It is not difficult to show that the Wasserstein distance provides a upper bound to the spectral distance [10, Prop. III.1],

$$d_D(\varphi, \tilde{\varphi}) \leq W_D(\varphi, \tilde{\varphi}) \quad \forall \varphi, \tilde{\varphi} \in \mathcal{S}(\mathcal{A}). \tag{14}$$

The equality holds on any subset of $\mathcal{S}(\mathcal{A})$ given by a convex linear combination of two pure states: fixed $\omega_1, \omega_2 \in \mathcal{P}(\mathcal{A})$, one denotes $\varphi_\lambda := \lambda\omega_1 + (1 - \lambda)\omega_2$. Then

$$d_D(\varphi_{\lambda_1}, \varphi_{\lambda_2}) = W_D(\varphi_{\lambda_1}, \varphi_{\lambda_2}) \quad \forall \lambda_1, \lambda_2 \in \mathbb{R}. \tag{15}$$

This shows in particular that $W_D = d_D$ on the whole of $\mathcal{S}(\mathcal{A})$ if $\mathcal{A} = M_2(\mathbb{C})$, since the pure state space of the algebra of 2×2 matrices is homeomorphic to the 2-sphere, so that the space of states is the 2-ball, and any two states φ_1, φ_2 can always be decomposed as two convex linear combinations $\varphi_{\lambda_1}, \varphi_{\lambda_2}$ of the same two pure states.

However the two distances are not equal in general, as can be seen in the following counter-example, taken from [13, §7]. Consider $\mathcal{A} = \mathbb{C}^3$ acting on $\mathcal{H} = \mathbb{C}^3$ as a diagonal matrix,

$$\pi(z_1, z_2, z_3) := \begin{pmatrix} z_1 & 0 & 0 \\ 0 & z_2 & 0 \\ 0 & 0 & z_3 \end{pmatrix} \quad \forall (z_1, z_2, z_3) \in \mathbb{C}^3, \tag{16}$$

[1] This may be true in general, but for safety we restrict to this well known case.

IOP Conf. Series: Journal of Physics: Conf. Series **968** (2018) 012007 doi:10.1088/1742-6596/968/1/012007

and take as Dirac operator

$$D = \begin{pmatrix} 0 & 0 & \alpha \\ 0 & 0 & \beta \\ \alpha & \beta & 0 \end{pmatrix} \qquad \alpha, \beta \in \mathbb{R}^+. \tag{17}$$

There are three pure states δ_i for $\mathcal{A}$, defined as

$$\delta_i(z_1, z_2, z_3) = z_i \quad i = 1, 2, 3. \tag{18}$$

So the space of states is the plain triangle with summit $\delta_1, \delta_2, \delta_3$.

By (15), one has that W_D coincides with d_D on each edge of the triangle. But the two distances do not agree on the whole triangle.

Proposition 3.1. *Let φ, φ' be states in $\mathcal{S}(\mathbb{C}^3)$,*

$$\varphi = \lambda_1\delta_1 + \lambda_2\delta_2 + (1 - \lambda_1 - \lambda_2)\delta_3, \qquad \varphi' = \lambda_1'\delta_1 + \lambda_2'\delta_2 + (1 - \lambda_1' - \lambda_2')\delta_3 \tag{19}$$

where $\lambda_i, \lambda_i' \in \mathbb{R}$, $i = 1, 2$ are such that $\Lambda_1 := \lambda_1 - \lambda_1'$ and $\Lambda_2 := \lambda_2 - \lambda_2'$ have the same sign. Then

$$W_D(\varphi, \varphi') = \frac{|\Lambda_1|}{\alpha} + \frac{|\Lambda_2|}{\beta} \tag{20}$$

while

$$d_D(\varphi, \varphi') = \sqrt{\frac{\Lambda_1^2}{\alpha^2} + \frac{\Lambda^2}{\beta^2}} \tag{21}$$

Proof. The cost function (10) is obtained computing explicitly the spectral distance between pure states (see e.g. [8, Prop. 7]):

$$d_D(\delta_1, \delta_2) = \sqrt{\frac{1}{\alpha^2} + \frac{1}{\beta^2}}, \qquad d_D(\delta_1, \delta_3) = \frac{1}{\alpha}, \qquad d_D(\delta_2, \delta_3) = \frac{1}{\beta}. \tag{22}$$

The Lipschitz ball (13) is thus

$$\mathrm{Lip}_D(\mathbb{C}^3) = \left\{ (z_1, z_2, z_3) \in \mathbb{C}^3 \,;\, |z_1 - z_2| \le \sqrt{\frac{1}{\alpha^2} + \frac{1}{\beta^2}},\, |z_1 - z_3| \le \frac{1}{\beta},\, |z_2 - z_3| \le \frac{1}{\alpha} \right\}. \tag{23}$$

For any $a = (z_1, z_2, z_3) \in \mathbb{C}^3$, one has

$$\varphi(a) - \varphi'(a) = \Lambda_1(z_1 - z_3) + \Lambda_2(z_2 - z_3). \tag{24}$$

Therefore

$$W_D(\varphi, \varphi') \le \frac{|\Lambda_1|}{\alpha} + \frac{|\Lambda_2|}{\beta}. \tag{25}$$

Since Λ_1 e Λ_2 have the same sign, this upper bound is attained by a_0 in $\mathrm{Lip}_D(\mathbb{C}^3)$ defined by

$$z_1 = \frac{1}{\alpha}, \qquad z_2 = \frac{1}{\beta}, \qquad z_3 = 0. \tag{26}$$

To prove (21), one computes the commutator of $a = (z_1, z_2, z_3)$ with D (see e.g. [8])

$$|||[D, a]||| = \sqrt{\alpha^2|z_3 - z_1|^2 + \beta^2|z_3 - z_2|^2}. \tag{27}$$

By subtracting $z_1\mathbb{I}$, one can always assume that $z_3 = 0$. The commutator condition $||[D, a]|| \leq 1$ becomes $\alpha^2|z_1|^2 + \beta^2|z_2|^2 \leq 1$, which is equivalent to

$$|z_2| \leq \sqrt{\frac{1-\alpha^2|z_1|^2}{\beta^2}}. \tag{28}$$

For such a, one obtains from (24)

$$|\varphi(a) - \varphi'(a)| \leq \Lambda_1|z_1| + \Lambda_2|z_2| \tag{29}$$

$$\leq \Lambda_1|z_1| + \Lambda_2\sqrt{\frac{1-\alpha^2|z_1|^2}{\beta^2}}. \tag{30}$$

On $[1, \frac{1}{\alpha}]$, the function $f(x) = \Lambda_1 x + \Lambda_2\sqrt{\frac{1-\alpha^2x^2}{\beta^2}}$ reaches its maximum when f' vanishes, that is for

$$x_0 = \frac{\Lambda_1}{\alpha^2\sqrt{\frac{\Lambda_1^2}{\alpha^2} + \frac{\Lambda_2^2}{\beta^2}}}. \tag{31}$$

This maximum,

$$f(x_0) = \sqrt{\frac{\Lambda_1^2}{\alpha^2} + \frac{\Lambda_2^2}{\beta^2}}, \tag{32}$$

is an upper bound for the spectral distance, reached by the element $a = (x_0, f(x_0), 0)$. □

4. Conclusion and outlook

On a manifold, Connes spectral distance between arbitrary states coincides with the Wasserstein distance of order 1 with cost the geodesic distance. On an arbitrary spectral triple, the spectral distance d_D on the space of states $\mathcal{S}(\mathcal{A})$ (viewed as the convex hull of the pure states space $\mathcal{P}(\mathcal{A})$) does not coincide with the Wasserstein distance W_D with cost function d_D (on $\mathcal{P}(\mathcal{A})$). However, W_D always provides an upper bound to the spectral distance, and the two distances coincides on any convex combination of two fixed pure states.

This shows that the interpretation of the spectral distance as a Wasserstein distance is more involved that could be initially thought, although the intriguing example worked out in this paper (W_D is the sum of the opposite and adjacent sides of a right a triangle, d_D is the length of the hypothenuse) suggests that there might exist a simple relation between the two distances.

Let us also mention that there do exists a formulation of the spectral distance as an infimum rather than a supremum (an analogue to "dual of the dual" formula of the Wasserstein distance in optimal transport), whose possible interpretation as a noncommutative cost still has to be elucidated [7].

References

[1] O. Bratteli and D. W. Robinson, *Operator algebras and quantum statistical mechanics 1*, Springer, 1987.
[2] A. Connes, *Gravity coupled with matter and the foundations of noncommutative geometry*, Commun. Math. Phys. **182** (1996), 155–176.
[3] A. Connes and J. Lott, *The metric aspect of noncommutative geometry*, Nato ASI series B Physics **295** (1992), 53–93.
[4] Alain Connes, *Noncommutative geometry*, Academic Press, 1994.
[5] Alain Connes, *On the spectral characterization of manifolds*, J. Noncom. Geom. **7** (2013), no. 1, 1–82.
[6] Francesco D'Andrea and Pierre Martinetti, *A view on optimal transport from noncommutative geometry*, SIGMA **6** (2010), no. 057, 24 pages.
[7] Francesco D'Andrea and Pierre Martinetti, in preparation.

IOP Conf. Series: Journal of Physics: Conf. Series **968** (2018) 012007 doi:10.1088/1742-6596/968/1/012007

[8] Bruno Iochum, Thomas Krajewski, and Pierre Martinetti, *Distances in finite spaces from noncommutative geometry*, J. Geom. Phy. **31** (2001), 100–125.

[9] L. V. Kantorovich, *On the transfer of masses*, Dokl. Akad. Nauk. SSSR **37** (1942), 227–229.

[10] P. Martinetti, *Towards a Monge-Kantorovich distance in noncommutative geometry*, Zap. Nauch. Semin. POMI **411** (2013).

[11] P. Martinetti, *From Monge to Higgs: a survey of distance computation in noncommutative geometry*, Contemporary Mathematics **676** (2016), 1–46.
[10]

[12] Marc A. Rieffel, *Metrics on states from actions of compact groups*, Documenta Math. **3** (1998), 215–229.

[13] Marc A. Rieffel, *Metric on state spaces*, Documenta Math. **4** (1999), 559–600.

IOP Conf. Series: Journal of Physics: Conf. Series **968** (2018) 012008 doi:10.1088/1742-6596/968/1/012008

Time functions and K-causality between measures

Tomasz Miller[1,2]

[1] Faculty of Mathematics and Information Science, Warsaw University of Technology, ul. Koszykowa 75, 00-662 Warsaw, Poland

[2] Copernicus Center for Interdisciplinary Studies, ul. Szczepańska 1/5, 31-011 Kraków, Poland

E-mail: `T.Miller@mini.pw.edu.pl`

Abstract. Employing the notion of a coupling between measures, drawn from the optimal transport theory, we study the extension of the Sorkin–Woolgar causal relation K^+ onto the space $\mathscr{P}(\mathcal{M})$ of Borel probability measures on a given spacetime $\mathcal{M}$. We show that Minguzzi's characterization of K^+ in terms of time functions possesses a "measure-theoretic" generalization. Moreover, we prove that the relation K^+ extended onto $\mathscr{P}(\mathcal{M})$ retains its property of antisymmetry for $\mathcal{M}$ stably causal.

1. Introduction

In causality theory, i.e. the subfield of mathematical relativity studying the causal properties of spacetimes, the fundamental role is played by the binary relations of chronological and causal precedence called I^+ and J^+, respectively. An event p is said to chronologically (causally) precede another event q if they can be connected by means of a future-directed timelike (causal) curve, what is usually denoted $p \ll q$ ($p \preceq q$). For a concise review of causality theory we refer the reader to [1], whereas a more detailed exposition can be found e.g. in [2, 3, 4, 5, 6].

There are, however, other binary relations studied in causality theory, which offer an alternative way of modelling causal influence. One of the important examples is the relation K^+, introduced by Sorkin and Woolgar [7] within the programme of reformulating causality theory along topological and order-theoretic lines, which in particular would encompass spacetimes with metrics of low regularity. K^+ was defined there as the smallest transitive and topologically closed relation containing I^+. Of course, if the spacetime $\mathcal{M}$ is causally simple, then J^+ is a closed subset of $\mathcal{M}^2$ and hence $K^+ = J^+$, but without the assumption of causal simplicity only the inclusion $K^+ \supseteq J^+$ holds, thus allowing for causal influence to involve more pairs of events than the causal curves alone can connect.

The Sorkin–Woolgar relation turns out to give some non-trivial insight into the property of *stable causality*, as attested by the following result due to Minguzzi [8] (see also the article by Bernard and Suhr in the present volume):

Theorem 1. *Spacetime $\mathcal{M}$ is stably causal iff it is K-causal, that is iff the relation K^+ is antisymmetric (and hence a partial order).*

In addition, it has been shown (also by Minguzzi [9]) that for $\mathcal{M}$ stably causal the relation K^+ can be completely characterized in terms of *time functions*, i.e. continuous real-valued maps on $\mathcal{M}$, which are strictly increasing along every future-directed causal curve. It is worth noting that such maps exist precisely on stably causal spacetimes [10]. Concretely, the characterization is as follows.

IOP Conf. Series: Journal of Physics: Conf. Series **968** (2018) 012008 doi:10.1088/1742-6596/968/1/012008

Theorem 2. *Let $\mathcal{M}$ be a stably causal spacetime and let $p, q \in \mathcal{M}$. Then*

$$(p, q) \in K^+ \quad \Leftrightarrow \quad \text{For every time function } t \quad t(p) \leq t(q). \tag{1}$$

Inspired by the causal structure in noncommutative Lorentzian geometry ([11], see also [12] and references therein) we have recently proposed a way to extend the causal precedence relation J^+ onto the space $\mathscr{P}(\mathcal{M})$ of all Borel probability measures on a given spacetime $\mathcal{M}$ [13]. The proposed extension employs the notion of a *coupling* of a pair of measures, drawn from the optimal transport theory [14, 15] adapted to the Lorentzian setting, and can be successfully used to model the causal time-evolution of spatially distributed quantities from classical physics, such as charge or energy densities [16, Section 2]. Surprisingly enough, it turns out to provide also a suitable framework for the study of quantum wave packets [17].

Let us emphasize, however, that the method of extending J^+ onto $\mathscr{P}(\mathcal{M})$ can be in fact applied to any (Borel) relation $R \subseteq \mathcal{M}^2$ (see Definition 1 below). In particular, in [18] we have studied the extension of the Sorkin–Woolgar relation K^+, addressing the question whether it retains its defining properties of transitivity and closedness, as well as establishing a "measure-theoretic" analogue of characterization (1). Although the first question has been answered positively, the characterization has been proven only under a stronger assumption of $\mathcal{M}$ being causally continuous. What is more, the problem whether K^+ extended onto $\mathscr{P}(\mathcal{M})$ is still antisymmetric for $\mathcal{M}$ stably causal has been left for future investigations.

The aim of the current paper is to fill both the above-mentioned gaps. To this end, we begin in Section 2 by providing the prerequisite definitions and some general results. Section 3 constitutes the main part of the article, with Theorem 5 answering the question concerning the characterization with time functions, and Theorem 7 tackling with the problem of antisymmetry. The (somewhat technical) proofs involve introducing and studying the so-called *multi-time orderings*, which generalize time orderings employed in [9].

2. Preliminaries

From now on, the term "measure" will always stand for "Borel probability measure".

Given a pair of measures $\mu, \nu \in \mathscr{P}(\mathcal{M})$, we call $\omega \in \mathscr{P}(\mathcal{M}^2)$ a *coupling* of μ and ν if the latter two measures are ω's marginals, that is if $\pi^1_\sharp \omega = \mu$ and $\pi^2_\sharp \omega = \nu$, where $\pi^i : \mathcal{M}^2 \to \mathcal{M}$, $i = 1, 2$ denote the canonical projections. We shall denote the set of all such couplings by $\Pi(\mu, \nu)$. Borrowing the terminology from Suhr [19, Definition 2.4], let us put forward the following general definition:

Definition 1. *Let $\mathcal{M}$ be a Polish space and let $R \subseteq \mathcal{M}^2$ be a* Borel *binary relation. For any $\mu, \nu \in \mathscr{P}(\mathcal{M})$ we say that μ is* R-related *with ν if there exists $\omega \in \Pi(\mu, \nu)$ such that $\omega(R) = 1$.*

Example 1. As a simple illustration of the above concept, consider the *equality relation* $=$, which, when regarded as a subset of $\mathcal{M}^2$, is nothing but $\Delta(\mathcal{M})$, where $\Delta : \mathcal{M} \to \mathcal{M}^2$, $\Delta(p) = (p, p)$ denotes the diagonal map. Then for any $\mu, \nu \in \mathscr{P}(\mathcal{M})$ the existence of $\omega \in \Pi(\mu, \nu)$ such that $\omega(\Delta(\mathcal{M})) = 1$ is a necessary and sufficient condition for μ and ν to be actually equal.

Indeed, necessity follows from the observation that, for any $\mu \in \mathscr{P}(\mathcal{M})$, $\Delta_\sharp \mu \in \Pi(\mu, \mu)$ and of course $\Delta_\sharp \mu(\Delta(\mathcal{M})) = \mu(\Delta^{-1}(\Delta(\mathcal{M}))) = \mu(\mathcal{M}) = 1$. For the proof of sufficiency, consult [13, Lemma 4]. □

By analogy with J^+, for any $\mathcal{X} \subseteq \mathcal{M}$ we introduce the notation $R^+(\mathcal{X}) := \pi^2((\mathcal{X} \times \mathcal{M}) \cap R)$ and $R^-(\mathcal{X}) := \pi^1((\mathcal{M} \times \mathcal{X}) \cap R)$. Notice that the sets $R^\pm(\mathcal{X})$ need not be Borel even if $\mathcal{X}$ is a Borel set. Nevertheless, being projections of Borel sets, they are *universally measurable*, which means that for any measure $\mu \in \mathscr{P}(\mathcal{M})$ the sets $R^\pm(\mathcal{X})$ are Borel up to a μ-negligible set, and therefore the quantity $\mu(R^\pm(\mathcal{X}))$ is well defined [20].

We will need the following powerful characterization of R-relatedness due to Suhr [19, Theorem 2.5].

IOP Conf. Series: Journal of Physics: Conf. Series **968** (2018) 012008 doi:10.1088/1742-6596/968/1/012008

Theorem 3. *Let $\mathcal{M}$ be a Polish space and let the relation $R \subseteq \mathcal{M}^2$ be closed (topologically). Then for any $\mu, \nu \in \mathscr{P}(\mathcal{M})$ the following conditions are equivalent*

(i) μ is R-related with ν.

(ii) For any Borel $\mathcal{B} \subseteq \mathcal{M}$

$$\mu(\mathcal{B}) \leq \nu(R^+(\mathcal{B})) \quad \text{and} \quad \mu(R^-(\mathcal{B})) \geq \nu(\mathcal{B}).$$

In this paper we will be dealing with *closed preorders*, i.e. those relations $R \subseteq \mathcal{M}^2$ which are reflexive, transitive and topologically closed. We will write $\mu \preceq_R \nu$ to express the R-relatedness of μ and ν. Notice that for any compact $C \subseteq \mathcal{M}$ the sets $R^{\pm}(C)$ are closed.

As a first general result, we obtain the following alternative characterizations of $\preceq_R$.

Theorem 4. *Let $\mathcal{M}$ be a Polish space and let $R \subseteq \mathcal{M}^2$ be a closed preorder. For any $\mu, \nu \in \mathscr{P}(\mathcal{M})$ the following conditions are equivalent:*

1• $\mu \preceq_R \nu$.

2• For any compact subset $C \subseteq \mathcal{M}$

$$\mu(R^+(C)) \leq \nu(R^+(C)). \tag{2}$$

3• For any Borel subset $\mathcal{X} \subseteq \mathcal{M}$ such that $R^+(\mathcal{X}) \subseteq \mathcal{X}$

$$\mu(\mathcal{X}) \leq \nu(\mathcal{X}). \tag{3}$$

Additionally, conditions 2• and 3• are equivalent with their "past" counterparts:

2′• For any compact subset $C \subseteq \mathcal{M}$

$$\mu(R^-(C)) \geq \nu(R^-(C)). \tag{4}$$

3′• For any Borel subset $\mathcal{Y} \subseteq \mathcal{M}$ such that $R^-(\mathcal{Y}) \subseteq \mathcal{Y}$

$$\mu(\mathcal{Y}) \geq \nu(\mathcal{Y}). \tag{5}$$

Proof. We adapt here of the first part of the proof of [18, Theorem 2].

1• $\Rightarrow$ 2• By the closedness of R, for any compact $C \subseteq \mathcal{M}$ the set $R^+(C)$ is closed and hence Borel. Denoting its characteristic function by χ (which is a Borel map), the inequality $\chi(p) \leq \chi(q)$ holds for all $(p,q) \in R$ by transitivity. Finally, on the strength of 1•, there exists $\omega \in \Pi(\mu, \nu)$ supported on R. Altogether, one can write that

$$\begin{aligned}\mu(R^+(C)) &= \int_{\mathcal{M}} \chi(p) d\mu(p) = \int_{\mathcal{M}^2} \chi(p) d\omega(p,q) = \int_{R} \chi(p) d\omega(p,q) \leq \int_{R} \chi(q) d\omega(p,q) \\ &= \int_{\mathcal{M}^2} \chi(q) d\omega(p,q) = \int_{\mathcal{M}} \chi(q) d\nu(q) = \nu(R^+(C)).\end{aligned}$$

One similarly proves that 1• $\Rightarrow$ 2′•.

2• $\Rightarrow$ 3• Let the set $\mathcal{X}$ be as specified in 3•. Taking any compact $C \subseteq \mathcal{X}$, we have that $R^+(C) \subseteq R^+(\mathcal{X}) \subseteq \mathcal{X}$ and hence

$$\mu(C) \leq \mu(R^+(C)) \leq \nu(R^+(C)) \leq \nu(\mathcal{X}),$$

where in the second inequality we have used $2^\bullet$. Using the fact that μ, being a Borel measure on a Polish space, is inner regular (a.k.a. *tight* [20, Lemma 12.6]), we hence obtain that

$$\mu(\mathcal{X}) = \sup\{\mu(C) \mid C \subseteq \mathcal{X} \text{ compact}\} \leq \nu(\mathcal{X}).$$

One similarly proves that $2'^\bullet \Rightarrow 3'^\bullet$.

$3^\bullet \Leftrightarrow 3'^\bullet$ Denote $\mathcal{X}^c := \mathcal{M} \setminus \mathcal{X}$. The possibility of moving between these two conditions relies on the obvious equality $\mu(\mathcal{X}^c) = 1 - \mu(\mathcal{X})$ valid for any $\mu \in \mathscr{P}(\mathcal{M})$ and any Borel set $\mathcal{X} \subseteq \mathcal{M}$, and on the following equivalence of inclusions: $R^+(\mathcal{X}) \subseteq \mathcal{X} \Leftrightarrow R^-(\mathcal{X}^c) \subseteq \mathcal{X}^c$. The latter has been stated in [13, Proposition 1] for J^+, but the proof conducted there is in fact valid for *any* relation R.

$3^\bullet, 3'^\bullet \Rightarrow 1^\bullet$ It suffices to notice that condition *(ii)* in Theorem 3 is satisfied.

Indeed, the first inequality in *(ii)* follows directly from $3^\bullet$ with $\mathcal{X} := R^+(\mathcal{B})$ (up to a μ-negligible set) and the obvious inequality $\mu(\mathcal{B}) \leq \mu(R^+(\mathcal{B}))$. Similarly, the second inequality follows from $3'^\bullet$ with $\mathcal{Y} := R^-(\mathcal{B})$ (up to a ν-negligible set) and another self-evident inequality $\nu(R^-(\mathcal{B})) \geq \nu(\mathcal{B})$. □

3. Main results

In [18], we have provided several characterizations of the K-causality relation between measures, some of which have been proven to hold for all stably causal spacetimes, whilst others seemed to demand a stronger requirement of causal continuity. Below, however, we show that the latter requirement is in fact redundant. In other words, we upgrade [18, Theorem 2] to the following result.

Theorem 5. *Let $\mathcal{M}$ be a stably causal spacetime. For any $\mu, \nu \in \mathscr{P}(\mathcal{M})$ the following conditions are equivalent.*

$1^\bullet$ $\mu \preceq_K \nu$.

$4^\bullet$ For any time function t and any $\lambda \in \mathbb{R}$

$$\mu\left(t^{-1}((\lambda, +\infty))\right) \leq \nu\left(t^{-1}((\lambda, +\infty))\right). \tag{6}$$

$5^\bullet$ For any bounded time function t

$$\int_{\mathcal{M}} t d\mu \leq \int_{\mathcal{M}} t d\nu. \tag{7}$$

The numbers $2^\bullet$, $3^\bullet$ have been omitted here as they refer to the conditions listed in Theorem 4, which of course holds in the special case $R := K^+$.

Additionally, the following two remarks from [18] are still valid (with their proofs unchanged).

Remark 1. Without loss of (or gain in) generality, in condition $5^\bullet$ the term "bounded" can be replaced with "μ- and ν-integrable", whereas the term "time" can be substituted with "temporal", "smooth time", "smooth causal" or "continuous causal".

Remark 2. Condition $4^\bullet$ can equivalently employ the *closed* half-lines. In other words, the following condition is equivalent to $4^\bullet$:

$4'^\bullet$ For any time function t and any $\lambda \in \mathbb{R}$

$$\mu\left(t^{-1}([\lambda, +\infty))\right) \leq \nu\left(t^{-1}([\lambda, +\infty))\right). \tag{8}$$

In order to prepare the ground for the proof of Theorem 5, we begin by slightly strengthening Minguzzi's Theorem 2.

IOP Conf. Series: Journal of Physics: Conf. Series **968** (2018) 012008 doi:10.1088/1742-6596/968/1/012008

Theorem 6. *Let $\mathcal{M}$ be a stably causal spacetime. Then there exists a* countable *family of* bounded *time functions $\{t_\alpha : \mathcal{M} \to (0,1)\}_{\alpha\in\mathbb{N}}$, such that for any $p,q \in \mathcal{M}$*

$$(p,q) \in K^+ \quad \Leftrightarrow \quad \forall\, \alpha \in \mathbb{N} \quad t_\alpha(p) \le t_\alpha(q) \tag{9}$$

and, moreover,

$$[\forall\, \alpha \in \mathbb{N} \quad t_\alpha(p) = t_\alpha(q)] \quad \Leftrightarrow \quad p = q. \tag{10}$$

Proof. The implication '$\Rightarrow$' in (9) follows trivially from Theorem 2. In order to show the converse, we will demonstrate how to construct a countable family of time functions $\{t_\alpha : \mathcal{M} \to (0,1)\}_{\alpha\in\mathbb{N}}$ such that, for any $p,q \in \mathcal{M}$

$$(p,q) \notin K^+ \quad \Rightarrow \quad \exists\, \alpha \in \mathbb{N} \quad t_\alpha(p) > t_\alpha(q).$$

To begin with, consider the open set $\mathcal{M}^2 \setminus K^+$. By Theorem 2, for any pair $(p,q) \in \mathcal{M}^2 \setminus K^+$ we can pick a time function $t_{p,q}$ such that $t_{p,q}(p) - t_{p,q}(q) > 0$. Notice now that the map $(t_{p,q} \circ \pi^1 - t_{p,q} \circ \pi^2) : \mathcal{M}^2 \to \mathbb{R}$ is continuous, and therefore the following family of inverse images:

$$\left\{(t_{p,q} \circ \pi^1 - t_{p,q} \circ \pi^2)^{-1}((0,+\infty))\right\}_{(p,q)\in M^2\setminus K^+}$$

constitutes an open cover of $\mathcal{M}^2 \setminus K^+$. Since the latter is a separable metric space (being an open subspace of a separable metric space $\mathcal{M}^2$), it possesses the Lindelöf property [21, Theorem 16.11], i.e. we can choose a countable subcover of the above cover:

$$\left\{(t_{p_\alpha,q_\alpha} \circ \pi^1 - t_{p_\alpha,q_\alpha} \circ \pi^2)^{-1}((0,+\infty))\right\}_{\alpha\in\mathbb{N}}.$$

To obtain the desired countable family of bounded time functions, it now suffices to define $t_\alpha := \varphi \circ t_{p_\alpha,q_\alpha}$ for every $\alpha \in \mathbb{N}$, where $\varphi \in C_b(\mathbb{R})$ is a fixed continuous strictly increasing function attaining values in $(0,1)$ (for instance, $\varphi(x) := \frac{1}{2} + \frac{1}{2}\tanh x$).

As for condition (10), it follows directly from (9) and Theorem 1. □

In [18] it was proven that the Sorkin–Woolgar relation retains its defining properties of transitivity and closedness when extended onto measures. With the help of Theorem 6, we can show that this extension is also antisymmetric (and hence a partial order) provided $\mathcal{M}$ is stably causal. In other words, Minguzzi's Theorem 1 still holds in the more general setting of K-causality between measures.

Theorem 7. *Let $\mathcal{M}$ be a stably causal spacetime. Then the relation $\preceq_K$ on $\mathscr{P}(\mathcal{M})$ is antisymmetric.*

Proof. The major part of the proof can be straightforwardly adapted from that of [13, Theorem 12], where it was proven that the relation J^+ (extended onto $\mathscr{P}(\mathcal{M})$) is antisymmetric under some mild assumptions on the causal properties of $\mathcal{M}$. In fact, there is only one step of that proof which requires a nontrivial modification. Namely, we need to show here that, for any fixed $\mu \in \mathscr{P}(\mathcal{M})$, the only $\omega \in \Pi(\mu,\mu)$ satisfying $\omega(K^+) = 1$ is $\Delta_\sharp\mu$. On the strength of [13, Lemma 4], it suffices to prove that $\omega(\Delta(\mathcal{M})) = 1$.

To this end, observe first that for any $f \in C_b(\mathcal{M})$ we have

$$\int_{K^+} (f(q) - f(p))\, d\omega(p,q) = 0, \tag{11}$$

IOP Conf. Series: Journal of Physics: Conf. Series **968** (2018) 012008 doi:10.1088/1742-6596/968/1/012008

what can be obtained by subtracting the identities $\int_{\mathcal{M}} f(p)d\mu(p) = \int_{K^+} f(p)d\omega(p,q)$ and $\int_{\mathcal{M}} f(q)d\mu(q) = \int_{K^+} f(q)d\omega(p,q)$, true by the very assumption on ω.

Let us now fix a countable family of time functions $\{t_\alpha : \mathcal{M} \to (0,1)\}_{\alpha\in\mathbb{N}}$ provided by Theorem 6. We claim that the following chain of equalities is true

$$\begin{aligned} 1 = \omega(K^+) &= \omega\left(\left\{(p,q) \in K^+ \,\Big|\, \sum_{\alpha=1}^{\infty} 2^{-\alpha}\, t_\alpha(p) = \sum_{\alpha=1}^{\infty} 2^{-\alpha}\, t_\alpha(q)\right\}\right) \\ &= \omega\left(\bigcap_{\alpha=1}^{\infty} \{(p,q) \in K^+ \mid t_\alpha(p) = t_\alpha(q)\}\right) = \omega(\Delta(\mathcal{M})). \end{aligned} \tag{12}$$

Indeed, the first equality follows from the very definition of ω. To see why the second equality holds, observe that the complementary subset $\{(p,q) \in K^+ \mid \sum_{\alpha=1}^{\infty} 2^{-\alpha}\, t_\alpha(p) < \sum_{\alpha=1}^{\infty} 2^{-\alpha}\, t_\alpha(q)\}$ is ω-null on the strength of (11), in which we have plugged $f := \sum_{\alpha=1}^{\infty} 2^{-\alpha}\, t_\alpha$. As for the third equality, one can actually prove the equality of the *sets* involved. In order to show the nontrivial inclusion "$\subseteq$", take any $(p,q) \in K^+$ such that $\exists\, \alpha_0 \in \mathbb{N}\ t_{\alpha_0}(p) < t_{\alpha_0}(q)$. Of course, since t_α's are all time functions, they satisfy $t_\alpha(p) \leq t_\alpha(q)$ by Theorem 1, and therefore we obtain that $\sum_{\alpha=1}^{\infty} 2^{-\alpha}\, t_\alpha(p) < \sum_{\alpha=1}^{\infty} 2^{-\alpha}\, t_\alpha(q)$. Finally, the fourth equality follows from condition (10). □

In the proof of Theorem 5 we will need another kind of causal relation, which generalizes Minguzzi's time orderings [9]. Recall that, given a time function $t : \mathcal{M} \to \mathbb{R}$, its corresponding *time ordering* is a (closed total) preorder defined as

$$T^+[t] = \{(p,q) \in \mathcal{M}^2 \mid t(p) \leq t(q)\}.$$

We generalize this definition onto finite collections of time functions in a straightforward way.

Definition 2. *Let $\mathcal{M}$ be a stably causal spacetime and let $F := \{t_1, t_2, \ldots, t_n\}$, $n \in \mathbb{N}$ be a finite set of time functions. Its corresponding* multi-time ordering *is a (closed) preorder defined via*

$$T^+[F] := \bigcap_{t\in F} T^+[t] = \{(p,q) \in \mathcal{M}^2 \mid t_\alpha(p) \leq t_\alpha(q),\ \alpha = 1,\ldots,n\}$$

Theorem 8. *Let $\mathcal{M}, F$ and n be as above. Then for any $\mu, \nu \in \mathscr{P}(\mathcal{M})$ the following conditions are equivalent:*

1• $\mu \preceq_{T[F]} \nu$.

2• *For any $\Phi \in C_b(\mathbb{R}^n)$ componentwise increasing*

$$\int_{\mathcal{M}} \Phi(t_1,\ldots,t_n)d\mu \leq \int_{\mathcal{M}} \Phi(t_1,\ldots,t_n)d\nu. \tag{13}$$

Proof. Implication 1• ⇒ 2• can be proven through the following chain of (in)equalities, valid for any componentwise increasing function $\Phi \in C_b(\mathbb{R}^n)$:

$$\begin{aligned} \int_{\mathcal{M}} \Phi(t_1,\ldots,t_n)d\mu &= \int_{K^+} \Phi(t_1(p),\ldots,t_n(p))d\omega(p,q) \\ &\leq \int_{K^+} \Phi(t_1(q),\ldots,t_n(q))d\omega(p,q) = \int_{\mathcal{M}} \Phi(t_1,\ldots,t_n)d\nu, \end{aligned}$$

where the equalities hold by the very definition of $\preceq_K$, whereas the inequality is true by Theorem 2 and the assumption on Φ.

IOP Conf. Series: Journal of Physics: Conf. Series **968** (2018) 012008 doi:10.1088/1742-6596/968/1/012008

In order to show the converse implication, we will demonstrate that $2^\bullet$ implies the following condition

$$\mu(T^+[F](C)) \leq \nu(T^+[F](C)) \quad \text{for any compact subset } C \subseteq \mathcal{M}, \tag{14}$$

which is equivalent to $1^\bullet$ by Theorem 4.

Let us first assume that C is *finite*. In order to show that inequality (14) holds for $C = \{q_1, \ldots, q_s\}$, consider the maps $\mathcal{T}_{k,l} : \mathcal{M} \to \mathbb{R}$ defined, for any $k, l \in \mathbb{N}$, via

$$\mathcal{T}_{k,l}(p) := \varphi_l^- \left(\sum_{i=1}^{s} \prod_{\alpha=1}^{n} \varphi_k^+ \left(t_\alpha(p) - t_\alpha(q_i)\right) \right),$$

where $\varphi_k^\pm \in C_b(\mathbb{R})$ are defined as $\varphi_k^\pm(x) := \frac{1}{2} + \frac{1}{2}\tanh(k^2 x \pm k)$. Observe that the sequence (φ_k^+) is pointwise convergent to the characteristic function of the *closed* half-line $\chi_{[0,+\infty)}$, whereas the pointwise limit of the sequence (φ_l^-) is the characteristic function of the *open* half-line $\chi_{(0,+\infty)}$. But this actually means that, for any $p \in \mathcal{M}$,

$$\lim_{l \to +\infty} \lim_{k \to +\infty} \mathcal{T}_{k,l}(p) = \begin{cases} 1 & \text{if } \exists i \in \{1, \ldots, s\}\ \forall \alpha \in \{1, \ldots, n\} \quad t_\alpha(p) \geq t_\alpha(q_i) \\ 0 & \text{otherwise.} \end{cases}$$

In other words, the iterated pointwise limit of the sequence $(\mathcal{T}_{k,l})$ calculated first with respect to k and then l is nothing but the characteristic function of the set $T^+[F](C)$.

Observe now that for any $k, l \in \mathbb{N}$ the map $\mathcal{T}_{k,l}$ is of the form $\Phi(t_1, \ldots, t_n)$ with $\Phi \in C_b(\mathbb{R}^n)$ componentwise increasing. On the strength of $2^\bullet$, we thus have that

$$\int_{\mathcal{M}} \mathcal{T}_{k,l}\, d\mu \leq \int_{\mathcal{M}} \mathcal{T}_{k,l}\, d\nu.$$

Invoking Lebesgue's dominated convergence theorem twice, we pass first to the limit $k \to +\infty$ and then to the limit $l \to +\infty$, thus obtaining (14) for C finite.

Suppose now that $C \subseteq \mathcal{M}$ is *any* compact subset. Our aim now is to construct a sequence $(C_m)_{m \in \mathbb{N}}$ of finite subsets of $\mathcal{M}$ such that

$$\forall m \in \mathbb{N} \quad T^+[F](C_{m+1}) \subseteq T^+[F](C_m) \quad \text{and} \quad \bigcap_{m=1}^{\infty} T^+[F](C_m) = T^+[F](C),$$

as this would already complete the proof of (14) via

$$\begin{aligned} \mu(T^+[F](C)) &= \mu\left(\bigcap_{m=1}^{\infty} T^+[F](C_m)\right) = \lim_{m \to +\infty} \mu\left(T^+[F](C_m)\right) \\ &\leq \lim_{m \to +\infty} \nu\left(T^+[F](C_m)\right) = \nu\left(\bigcap_{m=1}^{\infty} T^+[F](C_m)\right) = \nu(T^+[F](C)). \end{aligned}$$

We shall construct the sequence (C_m) recursively. To this end, let us first introduce the following open subsets: $B(C, R) := \bigcup_{x \in C} B(x, R)$ (for any $R > 0$) and, for any $p \in \mathcal{M}$,

$$V(p) := \bigcap_{\alpha=1}^{n} t_\alpha^{-1}\left((t_\alpha(p), +\infty)\right) = \{q \in \mathcal{M} \mid \forall \alpha \in \{1, \ldots, n\} \quad t_\alpha(p) < t_\alpha(q)\}.$$

IOP Conf. Series: Journal of Physics: Conf. Series **968** (2018) 012008 doi:10.1088/1742-6596/968/1/012008

Two immediate observations follow. First, for any $R > 0$ the family $\{V(p)\}_{p\in B(C,R)}$ constitutes an open cover of C (and hence it possesses a finite subcover). Indeed, for any $q \in C$ one can always choose $p \in B(q,R) \setminus \{q\}$ which causally precedes q and thus $t(p) < t(q)$ for any time function t. Second, it is easy to notice that $\overline{V(p)} = T^+[F](p)$.

To begin the construction of (C_m), let $R_1 := 1$ and let the family $\{V(p_1^{(1)}), \ldots, V(p_{s_1}^{(1)})\}$ be a finite subcover of the open cover $\{V(p)\}_{p\in B(C,R_1)}$ of C. Define $C_1 := \{p_1^{(1)}, p_2^{(1)}, \ldots, p_{s_1}^{(1)}\}$.

Suppose now we have constructed $C_m = \{p_1^{(m)}, \ldots, p_{s_m}^{(m)}\}$ for some $m \in \mathbb{N}$. In order to construct C_{m+1}, define first

$$R_{m+1} := \tfrac{1}{2} \min \left\{ \operatorname{dist} \left(C,\ \mathcal{M} \setminus \bigcup_{i=1}^{s_m} V(p_i^{(m)}) \right), R_m \right\},$$

where the distance (calculated with respect to some fixed auxiliary Riemannian metric on $\mathcal{M}$) is positive since C is compact and $\mathcal{M} \setminus \bigcup_{i=1}^{s_m} V(p_i^{(m)})$ is closed. Similarly as in the initial step, consider now a finite subcover $\{V(p_1^{(m+1)}), \ldots, V(p_{s_{m+1}}^{(m+1)})\}$ of the open cover $\{V(p)\}_{p\in B(C,R_{m+1})}$ of C and define $C_{m+1} := \{p_1^{(m+1)}, \ldots, p_{s_{m+1}}^{(m+1)}\}$.

Let us show that $T^+[F](C_{m+1}) \subseteq T^+[F](C_m)$ for any $m \in \mathbb{N}$. Indeed, take any $q \in T^+[F](C_{m+1})$, i.e. assume there exists $j \in \{1, \ldots, s_{m+1}\}$ such that $t_\alpha(p_j^{(m+1)}) \leq t_\alpha(q)$ for $\alpha = 1, \ldots, n$. Notice that, by construction,

$$p_j^{(m+1)} \in B(C, R_{m+1}) \subseteq \bigcup_{i=1}^{s_m} V(p_i^{(m)}) \subseteq \bigcup_{i=1}^{s_m} \overline{V(p_i^{(m)})} = T^+[F](C_m),$$

and therefore, altogether, there exists $i \in \{1, \ldots, s_m\}$ such that for any $\alpha \in \{1, \ldots, n\}$ $t_\alpha(p_i^{(m)}) \leq t_\alpha(p_j^{(m+1)}) \leq t_\alpha(q)$. Hence $q \in T^+[F](C_m)$.

Finally, it remains to prove that $\bigcap_{m=1}^{\infty} T^+[F](C_m) = T^+[F](C)$. In order to prove "$\subseteq$", take any $q \in \mathcal{M}$ satisfying

$$\forall m \in \mathbb{N}\ \ \exists p_m \in C_m\ \ \forall \alpha \in \{1, \ldots, n\}\quad t_\alpha(p_m) \leq t_\alpha(q).$$

By construction, for any $m \in \mathbb{N}$ one has $\operatorname{dist}(C, p_m) < R_m \leq 2^{m-1} \leq 1$. This implies that the sequence (p_m), being contained in a precompact set $B(C,1)$, has a subsequence convergent to some p, which actually must lie in $\bigcap_{m=1}^{\infty} B(C, 2^{m-1}) = \overline{C} = C$. By the continuity of time functions, this means that $t_\alpha(p) \leq t_\alpha(q)$ for $\alpha = 1, \ldots, n$ and so $q \in T^+[F](C)$.

To obtain the converse inclusion "$\supseteq$", fix any $m \in \mathbb{N}$ and recall that the family $\{V(p_i^{(m)})\}_{i=1}^{s_m}$ covers C, hence

$$C \subseteq \bigcup_{i=1}^{s_m} V(p_i^{(m)}) \subseteq \bigcup_{i=1}^{s_m} \overline{V(p_i^{(m)})} = T^+[F](C_m).$$

By transitivity, this implies that $T^+[F](C) \subseteq T^+[F](T^+[F](C_m)) = T^+[F](C_m)$. □

With the above characterization of multi-time orderings between measures at hand, we are finally ready to prove the characterization of $\preceq_K$ by means of time functions for $\mathcal{M}$ stably causal.

Proof of Theorem 5. Implications $1^\bullet \Rightarrow 4^\bullet \Rightarrow 5^\bullet$ have been established in [18], whereas the remaining implication $5^\bullet \Rightarrow 1^\bullet$ has been proven there only under the additional assumption of the causal continuity of $\mathcal{M}$. In order to show that this implication holds in all stably causal spacetimes, let us first fix a countable family of time functions $\{t_\alpha\}_{\alpha\in\mathbb{N}}$ with the property (9). On the strength of 5$^\bullet$ and Theorem 8 we obtain that for any $n \in \mathbb{N}$ there exists $\omega_n \in \Pi(\mu,\nu)$ such that

$$\omega_n\left(T^+[\{t_1,t_2,\ldots,t_n\}]\right) = \omega_n\left(\bigcap_{\alpha=1}^{n} T^+[t_\alpha]\right) = 1.$$

By the narrow compactness of $\Pi(\mu,\nu)$ (cf. [14, proof of Theorem 1.5]), the sequence $(\omega_n)_{n\in\mathbb{N}}$ has a subsequence $(\omega_{n_l})_{l\in\mathbb{N}}$ narrowly convergent to some $\omega \in \Pi(\mu,\nu)$. We claim that $\omega(K^+) = 1$, what would already yield 1$^\bullet$.

Indeed, observe first that for any $k,n \in \mathbb{N}$

$$k \leq n \quad \Rightarrow \quad \omega_n\left(\bigcap_{\alpha=1}^{k} T^+[t_\alpha]\right) \geq \omega_n\left(\bigcap_{\alpha=1}^{n} T^+[t_\alpha]\right) = 1.$$

Therefore, by the portmanteau theorem [22, Chapter 18, Theorem 6],

$$\forall\, k \in \mathbb{N} \quad 1 \geq \omega\left(\bigcap_{\alpha=1}^{k} T^+[t_\alpha]\right) \geq \limsup_{l\to+\infty} \omega_{n_l}\left(\bigcap_{\alpha=1}^{k} T^+[t_\alpha]\right) \geq 1,$$

and hence

$$\omega(K^+) = \omega\left(\bigcap_{\alpha=1}^{\infty} T^+[t_\alpha]\right) = \lim_{k\to+\infty} \omega\left(\bigcap_{\alpha=1}^{k} T^+[t_\alpha]\right) = 1,$$

where in the first equality we have used property (9). □

Acknowledgments

The author is grateful to Michał Eckstein for the careful perusal of the manuscript as well as to the Referees for their insightful comments. The research was made possible through the support of the National Science Centre, Poland via the grant Sonatina 2017/24/C/ST2/00322.

References

[1] Minguzzi E and Sánchez M 2008 *Recent developments in pseudo-Riemannian geometry, ESI Lectures in Mathematics and Physics* ed Alekseevsky D V and Baum H (Zürich: European Mathematical Society Publishing House) pp 299–358 (*Preprint* gr-qc/0609119)
[2] Beem J, Ehrlich P and Easley K 1996 *Global Lorentzian Geometry* (*Monographs and Textbooks in Pure and Applied Mathematics* vol 202) (New York: Marcel Dekker Inc.)
[3] Penrose R 1972 *Techniques of Differential Topology in Relativity* (*CBMS–NSF Regional Conference Series in Applied Mathematics* vol 7) (Philadelphia, PA: SIAM)
[4] O'Neill B 1983 *Semi-Riemannian Geometry with Applications to Relativity* (Cambridge, MA: Academic Press)
[5] Hawking S and Ellis G 1973 *The Large Scale Structure of Space-Time* (Cambridge: Cambridge University Press)
[6] Wald R 1984 *General Relativity* (Chicago, IL: University of Chicago Press)
[7] Sorkin R and Woolgar E 1996 *Class. Quantum Grav.* **13** 1971–93
[8] Minguzzi E 2009 *Commun. Math. Phys.* **290** 239–48 (*Preprint* gr-qc/0809.1214)
[9] Minguzzi E 2010 *Commun. Math. Phys.* **298** 855–68 (*Preprint* gr-qc/0909.0890)
[10] Sánchez M 2005 *Matemática Contemporânea* **29** 127–55 (*Preprint* gr-qc/0411143)
[11] Franco N and Eckstein M 2013 *Class. Quantum Grav.* **30** 135007 (*Preprint* math-ph/1212.5171)
[12] Eckstein M 2017 *Universe* **3** 25

[13] Eckstein M and Miller T 2017 *Ann. Henri Poincaré* **18** 3049–96
[14] Ambrosio L and Gigli N 2013 *Modelling and Optimisation of Flows on Networks: Cetraro, Italy 2009* ed Piccoli B and Rascle M (Berlin, Heidelberg: Springer) pp 1–155
[15] Villani C 2008 *Optimal Transport: Old and New* (*Grundlehren der mathematischen Wissenschaften* vol 338) (Berlin, Heidelberg: Springer)
[16] Miller T 2017 *J. Geom. Phys.* **116** 295–315
[17] Eckstein M and Miller T 2017 *Phys. Rev. A* **95**(3) 032106
[18] Miller T 2017 *Universe* **3** 27
[19] Suhr S 2016 Theory of optimal transport for Lorentzian cost functions *Preprint* 1601.04532 [math.DG]
[20] Aliprantis C and Border K 2006 *Infinite Dimensional Analysis: A Hitchhiker's Guide* (Berlin, Heidelberg: Springer)
[21] Willard S 1970 *General Topology* (Reading, MA: Addison-Wesley)
[22] Fristedt B and Gray L 1997 *A Modern Approach to Probability Theory* (Basel: Birkhäuser)

IOP Conf. Series: Journal of Physics: Conf. Series **968** (2018) 012009 doi:10.1088/1742-6596/968/1/012009

The representation of spacetime through steep time functions

Ettore Minguzzi
Dipartimento di Matematica e Informatica "U. Dini", Università degli Studi di Firenze, Via S. Marta 3, I-50139 Firenze, Italy
E-mail: ettore.minguzzi@unifi.it

Abstract. In a recent work I showed that the family of smooth steep time functions can be used to recover the order, the topology and the (Lorentz-Finsler) distance of spacetime. In this work I present the main ideas entering the proof of the (smooth) distance formula, particularly the product trick which converts metric statements into causal ones. The paper ends with a second proof of the distance formula valid for globally hyperbolic Lorentzian spacetimes.

1. Introduction

In a recent work [1] I obtained optimal conditions for the existence of steep time functions on spacetime. The result was used to characterize the Lorentzian submanifolds of Minkowski spacetime and to prove the (smooth Lorentz-Finsler) distance formula. At the meeting I announced the latter result while placing it into the broad context of functional representation results for topological ordered spaces [2,3]. In this work I shall outline the proof strategy instead, by making use of some illustrations, and by leaving the technical details to the original paper.

Unless otherwise stated, we shall work in the general context of closed cone structures and closed Lorentz-Finsler spaces. Let M be a connected, Hausdorff, second-countable C^1 manifold. A *closed cone structure* (M, C), $C \subset TM\backslash 0$, is a closed cone subbundle of the slit tangent bundle such that $C_x := C \cap T_xM$ is a closed (in the topology of $T_xM\backslash 0$), sharp, convex, non-empty cone. The multivalued map $x \mapsto C_x$ turns out to be upper semi-continuous, that is as y approaches x, using the identification of tangent spaces provided by any coordinate system, $C_y \backslash C_x \to \emptyset$, cf. [1, 4] for a more precise definition of the last limit. Other useful differentiability conditions on the multivalued map are continuity and local Lipschitzness [1, 5]. We stress that our cones C_x are not necessarily strictly convex, a feature that will be important, as we shall see.

The closed cone structure conveys the notion of causality. The *causal vectors* are the elements of C, the *timelike vectors* are the elements of Int$C$ while those belonging to $C\backslash$Int$C$ might be called *lightlike vectors*. In the previous expressions Int is the interior for the topology of $TM\backslash 0$. It can be noticed that $(\mathrm{Int}C)_x \subset \mathrm{Int}C_x$ and equality holds for C^0 cone structures. A continuous causal curve is an absolutely continuous map $x : I \to M$ which has causal derivative almost everywhere. The causal relation J consists of all those pairs of events (p, q) such that there is a continuous causal curve connecting p to q or $p = q$. *Closed* cone structures are particularly well behaved since for them the limit curve theorem holds true. Actually, several other non trivial results hold true for closed cone structures, such as the causal ladder of spacetimes. We shall not explore these findings in this work; the reader is referred to [1] for more details.

IOP Conf. Series: Journal of Physics: Conf. Series **968** (2018) 012009 doi:10.1088/1742-6596/968/1/012009

The length of vectors is measured by the fundamental Lorentz-Finsler function $\mathscr{F}: C \to [0, +\infty)$ which is positive homogeneous and concave. As a consequence, it satisfies the reverse triangle inequality: for every $y_1, y_2 \in C$, $\mathscr{F}(y_1 + y_2) \geq \mathscr{F}(y_1) + \mathscr{F}(y_2)$. In this work a *Lorentz-Finsler space* is just a pair $(M, \mathscr{F})$, otherwise called a *spacetime*.

Lorentzian geometry is obtained with the choice

$$\mathscr{F}(y) = \sqrt{-g(y,y)}, \tag{1}$$

where g is the Lorentzian metric. Its cones C_x are said to be *round* since the intersection with an affine plane on T_xM is an ellipsoid.

It must be mentioned that in standard Lorentz-Finsler theories the Finsler Lagrangian $\mathscr{L} = -\frac{1}{2}\mathscr{F}^2$ has Lorentzian vertical Hessian g on C, $\mathscr{F}(\partial C) = 0$, and further regularity properties are demanded for $\mathscr{L}$, e.g. g is continuous up to ∂C. The Lorentz-Finsler spaces of this work are much more general, for instance we do not even assume $\mathscr{F}(\partial C) = 0$. However, we are left with the problem of introducing convenient regularity conditions on $\mathscr{F}$.

One of the key ideas of our work concerns the very definition of *closed Lorentz-Finsler space*. This is one instance of application of the *product trick*. We observe that $\mathscr{F}$ with its properties allows us to define a sharp convex non-empty cone at every tangent space of $M^\times = M \times \mathbb{R}$, which we give in two versions

$$C^\times = \{(y,z) \colon y \in C \cup \{0\}, |z| \leq \mathscr{F}(y)\} \backslash \{0,0\} \tag{2}$$

$$C^\downarrow = \{(y,z) \colon y \in C \cup \{0\}, z \leq \mathscr{F}(y)\} \backslash \{0,0\} \tag{3}$$

The former version would seem the most natural since in Lorentzian geometry it produces a round cone, but it is actually the latter which will prove to be the most useful. Observe that $C^\downarrow$ is not strictly convex.

The main idea is that the cone structure $(M^\times, C^\downarrow)$ encodes all the information on the Lorentz-Finsler space $(M, \mathscr{F})$ so in order to get the best results for $(M, \mathscr{F})$ we have to impose those differentiability conditions on $\mathscr{F}$ which guarantee that $(M^\times, C^\downarrow)$ is a nice type of cone structure. So we impose that $(M^\times, C^\downarrow)$ is a closed cone structure, a condition which is equivalent to the upper semi-continuity of both C and $\mathscr{F}$. By definition, with these conditions $(M, \mathscr{F})$ is a *closed Lorentz-Finsler space*. At this point, instead of working out new results for Lorentz-Finsler spaces we can just translate the known results for cone structures. We shall follow this approach in the construction of steep time functions, but in the main paper one can find this idea applied in many directions, from the definition of causal geodesic to the proof of the notable singularity theorems.

A *time function* is a continuous function which increases over every causal curve. A *temporal function* is a C^1 function t such that $\mathrm{d}t(y) > 0$ for every $y \in C$, hence a time function. Let $f \colon C \to [0, +\infty)$ be positive homogeneous. An f-steep function is a C^1 function t such that

$$\mathrm{d}t(y) \geq f(y) \tag{4}$$

for every $y \in C$. It is *strictly* f-steep if the inequality is strict. Since $\mathscr{F} \geq 0$, every strictly $\mathscr{F}$-steep function is temporal. Our main objective is to characterize those closed Lorentz-Finsler spaces which admit a strictly $\mathscr{F}$-steep function. If h is a Riemannian metric, with some abuse of terminology, we say that a function is h-steep if it is $|\cdot|_h$-steep.

The (Lorentz-Finsler) length of a continuous causal curve $x \colon [0,1] \to M$, $t \mapsto x(t)$, is

$$\ell(x) = \int_0^1 \mathscr{F}(\dot{x})\mathrm{d}t$$

IOP Conf. Series: Journal of Physics: Conf. Series **968** (2018) 012009 doi:10.1088/1742-6596/968/1/012009

(it is independent of the parametrization). The (Lorentz-Finsler) distance is defined by: for $(p,q)\notin J$, $d(p,q)=0$, while for $(p,q)\in J$

$$d(p,q)=\sup_x \ell(x), \tag{5}$$

where x runs over the continuous causal curves which connect p to q.

It can be observed that if $x\colon [0,1]\to M$ is continuous causal then $x^\times : I\to M^\times$ given by

$$x^\times(t)=(x(t),\int_0^t \mathscr{F}(\dot x(s))\mathrm{d}s)$$

is continuous causal. Let us set $x(0)=p$, $x(1)=q$, $P=(p,0)$, $Q=(q,\ell(x))$ then Q is the endpoint of the continuous causal curve $x^\times$. Thus $d(p,q)$ is really an upper bound for the fiber coordinate over $(J^\downarrow)^+(P)\cap\pi^{-1}(q)$ where $J^\downarrow$ is the causal relation on $(M,C^\downarrow)$, and π is the projection $M^\times\to M$. Notice that $d(p,q)$ would be the maximum if $J^\downarrow$ were a closed relation.

We write $C'>C$ if C' is a C^0 cone structure such that $C\subset \mathrm{Int}C'$. We say that C is a *causal* closed cone structure if it does not admit closed continuous causal curves. A *stably causal* closed cone structure is one for which we can find $C'>C$ such that C' is causal. It is by now well established that under stable causality the most useful causal relation is the Seifert relation

$$J_S=\bigcap_{C'>C} J'$$

since it is closed and transitive. Under stronger causality conditions, such as causal simplicity or global hyperbolicity J is closed, a fact which implies that $J_S=J$. In fact, under stable causality J_S is really the smallest closed and transitive relation which contains J. This result, conjectured by Low in 1996, and implicit in some of Seifert's works in the early seventies was proved by the author in [6] at least for the C^2 Lorentzian theory. However, the proof was really topological so we could generalize it to cone structures [1]. It must be said that in [1] the proof appears after the construction of the steep time functions, thus the order of presentation is reversed with respect to this work. In fact in that work we looked for the most convenient proofs while here our goal is just that of showing that our steep time function construction is natural and reasonable given the experience built on Lorentzian geometry.

The cone structure $C^\downarrow$ is not round, in fact it is not even strictly convex, but it is this cone structure that will prove to be fundamental for our arguments. So it is natural to consider causality theory for non-round cone structures even if one's interest is in Lorentzian geometry.

Among the results which are known to hold in Lorentzian geometry [7] and which survive in the cone structure case [1,8] we can find the following: in a stably causal spacetime the Seifert relation can be represented with the set of smooth temporal functions, that is

$$(p,q)\in J_S \Leftrightarrow t(p)\le t(q) \quad \text{for every smooth temporal function } t. \tag{6}$$

Let us return to the product trick. We know that a closed Lorentz-Finsler space is best seen as a cone structure on $M^\times$, so it is natural to ask what happens after slightly opening the cones on $M^\times$. What is the geometrical meaning of $J_S^\downarrow$, the Seifert relation on $(M^\times, C^\downarrow)$? We have seen that $d(p,q)$ is not the maximum of the fiber coordinate on $(J^\downarrow)^+(P)\cap\pi^{-1}(q)$ just because $J^\downarrow$ is not closed. But $J_S^\downarrow$ is, and moreover it is contained in $J'^\downarrow$ for every opening of the cones $C'^\downarrow > C^\downarrow$. Here this opening implies the enlargement of the cone, so $C'>C$, and that of the graphing function $\mathscr{F}$ thus $\mathscr{F}'>\mathscr{F}$. Actually, as a matter of notation, with the latter inequality we shall always include the validity of the former.

It will therefore not come as a surprise that

$$J_S^\downarrow=\{((p,r),(p',r'))\colon (p,p')\in J_S \text{ and } r'-r\le D(p,p')\}. \tag{7}$$

IOP Conf. Series: Journal of Physics: Conf. Series **968** (2018) 012009 doi:10.1088/1742-6596/968/1/012009

where the *stable distance* $D\colon M \times M \to [0, +\infty]$ is defined as follows. For $p, q \in M$

$$D(p, q) = \inf_{\mathscr{F}' > \mathscr{F}} d'(p, q), \tag{8}$$

where d' is the Lorentz-Finsler distance for the Lorentz-Finsler space $(M, \mathscr{F}')$.

The closure and transitivity of $J_S^{\downarrow}$ are reflected in two important properties of D, namely upper semi-continuity and reverse triangle inequality: for every $(p, q) \in J_S$ and $(q, r) \in J_S$

$$D(p, r) \geq D(p, q) + D(q, r).$$

It can be noticed that D is better behaved than d since the reverse triangle inequality has wider applicability. It can be shown that for globally hyperbolic cone structures $D = d$, but for less demanding causality conditions D, rather than d, should be regarded as the most convenient Lorentz-Finsler distance.

We say that a closed Lorentz-Finsler space is *stable* if it is stable *causally* and *metrically*, namely if it is stably causal and stably finite. By *stably finite* we mean that d remains finite under small perturbations of the Finsler function $\mathscr{F}$, namely there is $\mathscr{F}' > \mathscr{F}$ such that d' is finite. Clearly, this condition implies that $D < +\infty$, but one of the main result of our work, a bit lengthy to be discussed here, states that the converse is true: if $D < +\infty$ then d is stably finite [1].

Finally, we mention that globally hyperbolic spacetimes are stable no matter the choice of $\mathscr{F}$ while stably causal spacetimes are conformally stable.

2. The distance formula

Our goal is to prove the distance formula. Alain Connes in the early nineties proposed a strategy for the unification of all fundamental physical forces based on non-commutative geometry [9]. A key ingredient was the *distance formula*, an expression for the Riemannian distance in terms of the 1-Lipschitz functions on spacetime. Unfortunately, the signature of the metric on spacetime is Lorentzian so Connes did not describe correctly the spacetime manifold: there was no causality or Lorentzian distance.

Parfionov and Zapatrin [10] proposed to consider a more physical Lorentzian version and for that purpose they introduced the notion of steep time function which we already met. Let d denote the Lorentzian distance, and let $\mathscr{S}$ be the family of $\mathscr{F}$-steep time functions. The Lorentzian version of Connes' distance formula would be, for every $p, q \in M$

$$d(p, q) = \inf\{[f(q) - f(p)]^+ \colon\ f \in \mathscr{S}\}. \tag{9}$$

where $c^+ = \max\{0, c\}$.

Progress in the proof of the formula was made by Moretti [11] and Franco [12] for globally hyperbolic spacetimes. However, in their versions the functions appearing in (9) were not differentiable everywhere a fact which was a little annoying since in Connes' program the Dirac operator acts on them. We proved the smooth version for stable spacetimes as part of the next more general theorem on the representability of spacetime by steep time function [1].

Theorem 2.1. *Let $(M, \mathscr{F})$ be a closed Lorentz-Finsler space and let $\mathscr{S}$ be the family of smooth strictly $\mathscr{F}$-steep temporal functions. The Lorentz-Finsler space $(M, \mathscr{F})$ is stable if and only if $\mathscr{S}$ is non-empty. In this case $\mathscr{S}$ represents*

(a) the order J_S, namely $(p, q) \in J_S \Leftrightarrow f(p) \leq f(q),\ \forall f \in \mathscr{S}$;

(b) the manifold topology, namely for every open set $O \ni p$ we can find $f, h \in \mathscr{S}$ in such a way that $p \in \{q\colon f(q) > 0\} \cap \{q\colon h(q) < 0\} \subset O$;

IOP Conf. Series: Journal of Physics: Conf. Series **968** (2018) 012009 doi:10.1088/1742-6596/968/1/012009

(c) the stable distance, in the sense that the distance formula holds true: for every $p, q \in M$

$$D(p,q) = \inf\{[f(q) - f(p)]^+ : \ f \in \mathscr{S}\}. \tag{10}$$

Moreover, strictly *can be dropped.*

As said above we shall concentrate on the distance formula, i.e. point (c). We have also a version for stably causal rather than stable spacetimes. However, in that version the steep functions have to be taken with codomain $[-\infty, +\infty]$ and steep just on the finite set.

Here the main idea is very simple and follows from Eq. (7): the distance formula is a consequence of (6) applied to $(M^\times, C^\downarrow)$, thus as a first step we have to show that this product spacetime is stably causal (we shall return to it later on). At this point the level sets of temporal functions on $(M^\times, C^\downarrow)$ are really local graphs of strictly $\mathscr{F}$-steep functions on M. It is at this step that the shape of the cone $C^\downarrow$ becomes essential, since it is thanks to the fact that $C^\downarrow$ contains the fiber direction that the level set is a (univalued) graph, see Figure 1.

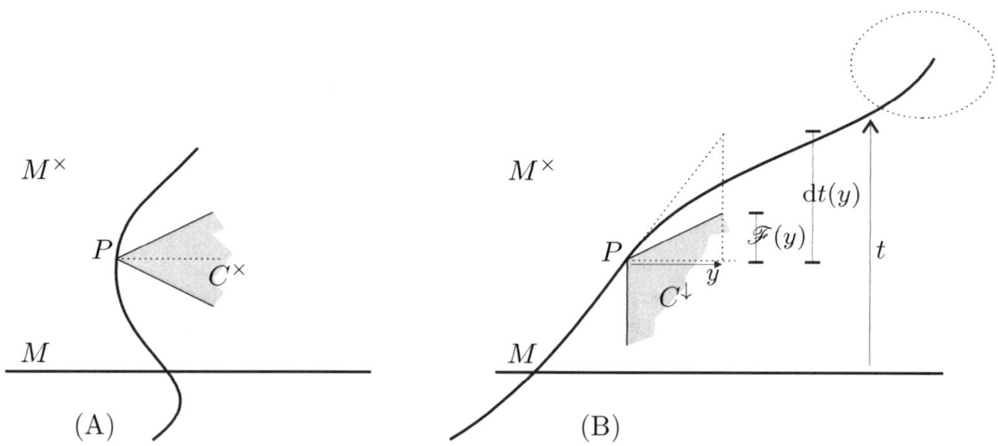

Figure 1. The level set of a temporal function τ on $(M^\times, C^\times)$ and on $(M^\times, C^\downarrow)$. In the latter case, due to the shape of the cone, it is necessarily the local graph of a function t which, furthermore, is strictly $\mathscr{F}$-steep: for every $y \in C$, $\mathrm{d}t(y) > \mathscr{F}(y)$. The domain of t might not be the whole M since t might blow up. This problem is resolved by constructing τ directly through a modification of Hawking's averaging method and by using the condition $D < +\infty$.

Let us see what are the implications of (6) on $(M^\times, C^\downarrow)$. Let $P = (p, r)$ be a generic point, let $q \in J^+_S(p)$ and let $Q = (q, r') = (q, r + D(p, q))$. By Eq. (7) $(P, Q) \in J^\downarrow_S$. By (6) if τ is temporal on $(M, C^\downarrow)$, $\tau(P) \le \tau(Q)$ which means that Q must stay in the subgraph of the level set of τ passing through P. In other words $t(q) - t(p) \ge r' - r = D(p, q)$, see Fig. 2. But for every strictly $\mathscr{F}$-steep function $t\colon M \to \mathbb{R}$ we can find a temporal function $\tau(P) = t(p) - r$ which has the graph of t as level set. Thus the previous argument shows that for every strictly $\mathscr{F}$-steep function t, $D(p, q) \le t(q) - t(p)$. We now show that the infimum of the right-hand side is really D, thus concluding the proof. In fact (6) states that the temporal functions separate points not belonging to the Seifert relation. Let $Q_\epsilon = (q, r + D(p, q) + \epsilon)$, for $\epsilon > 0$, then by Eq. (7) $(P, Q_\epsilon) \notin J^\downarrow_S$. As a consequence, there is a temporal function τ on $(M^\times, C^\downarrow)$ which separates them in the sense that $\tau(Q_\epsilon) < \tau(P)$, which implies that the graphing function of the τ-level set passing through P satisfies $t(q) - t(p) \le D(p, q) + \epsilon$, see Fig. 2. Due to the arbitrariness of ϵ we get the distance formula for $(p, q) \in J_S$.

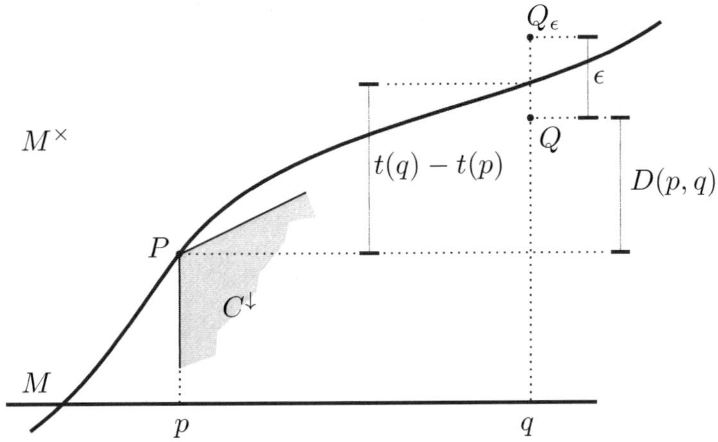

Figure 2. Since $(P, Q_\epsilon) \notin J_S^\downarrow$, there is a temporal function $\tau\colon M^\times \to \mathbb{R}$ which separates P and Q_ϵ in the sense that $\tau(Q_\epsilon) < \tau(P)$. The figure displays the level set of τ passing though P.

Actually, the function t might not be defined everywhere but it turns out that it is sufficient that there exists *one* (global) strict $\mathscr{F}$-steep function to prove that such functions separate points not belonging to $J_S^\downarrow$. So we are left with two problems

(i) We have yet to prove that $(M, C^\downarrow)$ is stably causal, a fact assumed when we made use of property (6) on $(M^\times, C^\downarrow)$. This result can be proved constructing directly a time function τ on $(M, C^\downarrow)$. In fact, we can do this with a sort of Hawking's averaging method.

(ii) We have to show that this function τ, suitably smoothed, gives a temporal function on $(M, C^\downarrow)$ whose level sets intersects every $\mathbb{R}$-fiber (so that the graphing function t of a level set is globally defined).

While (i) can be accomplished under stable causality of (M, C), (ii) holds only if (M, C) is stable. In fact, we know that the inequality $D(p, q) \le t(p) - t(q)$ holds for every strictly $\mathscr{F}$-steep function t, thus the existence of just one such function implies that D is finite.

Concerning step (i) we recall that Hawking's averaging method [13, 14] applied to $(M^\times, C^\downarrow)$ would consist in the introduction of a positive normalized measure μ on $M^\times$, absolutely continuous with respect to the Lebesgue measure of any chart, in the introduction of a one-parameter family of cones $C_a^\downarrow$, $a \in [0, 3]$, $C_a^\downarrow < C_b^\downarrow$, for $a < b$, $C_0^\downarrow = C^\downarrow$, and in the definition of

$$\tau^\downarrow(P) = -\int_1^2 \mu\big(J^+_{C_a^\downarrow}(P)\big)\mathrm{d}a.$$

This construction gives a time function on $(M^\times, C^\downarrow)$ provided this spacetime is stably causal. In fact one would have to choose $C_3^\downarrow$ (stably) causal. Unfortunately, we do not know if $C^\downarrow$ is stably causal. Instead, we construct the function $\tau^\downarrow$ using a different choice for $C_a^\downarrow$ which is causal by construction, namely $C_a^\downarrow$ is built from $(C_a, \mathscr{F}_a)$ according to the analog of Eq. (3) where the one-parameter family $(C_a, \mathscr{F}_a)$ is defined for $a \in [0, 3]$, and satisfies: $C_a < C_b$ and $\mathscr{F}_a < \mathscr{F}_b$, for $a < b$, $C_0 = C$ and $\mathscr{F}_0 = \mathscr{F}$, C_3 is stably causal. In other words, the cones $C_a^\downarrow$ do not open in the direction of the $\mathbb{R}$-fiber which for all of them remains lightlike. They are causal because with this definition the projection of a continuous $C_a^\downarrow$-causal curve is a continuous C_a-causal curve (unless it coincides with a fiber), and so there cannot be a closed causal curve on $M^\times$ as there would be one on the base, which is impossible as C_3 is stably causal.

IOP Conf. Series: Journal of Physics: Conf. Series **968** (2018) 012009 doi:10.1088/1742-6596/968/1/012009

With this definition $\tau^{\downarrow}$ is clearly increasing over continuous $C^{\downarrow}$-causal curves since the argument of the integral is for every a. What is puzzling is the fact that $\tau^{\downarrow}$ is continuous despite the fact that the cones are not opened as in Hawking's original prescription. This point cannot be understood intuitively but it is a consequence of two facts: (a) the invariance under the fiber translations of the cone structure $C^{\downarrow}$, (b) the fact that $C^{\downarrow}$ projects to a *sharp* cone C. Notice that the latter property would not hold if $C^{\downarrow}$ were round (Lorentzian), since the projection would be a half-space.

Coming to step (ii) the function τ can be smoothed thanks to a powerful result [1] which improves a previous result by Chruściel, Grant and the author [15, Th. 4.8]. I recall the theorem though I shall not enter into the details of its use (one can apply it directly to the function τ on $M^{\times}$ or to the graphing function t on M).

Theorem 2.2. *Let (M, C) be a closed cone structure and let $\tau\colon M \to \mathbb{R}$ be a continuous function. Suppose that there is a C^0 proper cone structure $\hat{C} > C$ and continuous functions homogeneous of degree one in the fiber $\underline{F}, \overline{F}\colon \hat{C} \to \mathbb{R}$ such that for every $\hat{C}$-timelike curve $x\colon [0,1] \to M$*

$$\int_x \underline{F}(\dot{x})\mathrm{d}t \le \tau(x(1)) - \tau(x(0)) \le \int_x \overline{F}(\dot{x})\mathrm{d}t. \tag{11}$$

Let h be an arbitrary Riemannian metric, then for every function $\alpha\colon M \to (0, +\infty)$ there exists a smooth function $\hat{\tau}$ such that $|\hat{\tau} - \tau| < \alpha$ and for every $v \in C$

$$\underline{F}(v) - \|v\|_h \le \mathrm{d}\hat{\tau}(v) \le \overline{F}(v) + \|v\|_h. \tag{12}$$

Similar versions, in which some of the functions $\underline{F}, \overline{F}$ do not exist hold true. One has just to drop the corresponding inequalities in (12).

More interesting, and peculiar to the proof, is how we solved the problem of constructing τ in such a way that its level sets intersect every $\mathbb{R}$-fiber. Here the idea comes from Geroch's original construction of the time function on globally hyperbolic spacetime [14]. Basically, we repeat the construction above with the opposite cones, thus obtaining a time function

$$\tau^{\uparrow}(P) = \int_1^2 \mu(J^-_{C_a^{\downarrow}}(P))\mathrm{d}a.$$

Then we define $\tau = \log|\tau^{\uparrow}/\tau^{\downarrow}|$. If we can show that $\tau^{\downarrow} \to 0$ moving to the future along a fiber (the downward direction in our figures) and that $\tau^{\uparrow} \to 0$ moving to the past along a fiber (the upward direction in our figures) then $\tau \to +\infty$ in the former limit and $\tau \to -\infty$ in the latter, thus by continuity the level set $\tau = 0$ intersects every fiber. The condition $D < +\infty$ guarantees precisely the mentioned limits. Indeed if $K \subset M$ is a compact set then $J^+_{C_a^{\downarrow}}(P) \cap \pi^{-1}(K)$ will be upper bounded in the extra coordinate, with bound going to $-\infty$ as the fiber coordinate of P goes to $-\infty$. This fact should not totally come as a surprise since as we learned in Fig. 2, the function D controls how much we can go up in the extra-coordinate with respect to the starting point.

Our discussion of points (i) and (ii) finishes our exposition of the proof that under the stable condition $D < +\infty$ there are (global) strictly $\mathscr{F}$-steep time functions, which was the missing step in our argument for proving the distance formula.

Incidentally the existence of $\mathscr{F}$-steep functions is important in the problem of embedding Lorentzian manifolds into Minkowski spacetime, as shown by Müller and Sánchez [16]. As a consequence we also proved a long sought for characterization of the Lorentzian submanifolds of Minkowski spacetime. According to it the Lorentzian submanifolds are precisely the stable spacetimes [1], or more precisely

IOP Conf. Series: Journal of Physics: Conf. Series **968** (2018) 012009 doi:10.1088/1742-6596/968/1/012009

Theorem 2.3. *Let (M, g) be a $n+1$-dimensional Lorentzian spacetime endowed with a C^k, $3 \le k \le \infty$, metric. (M, g) admits a C^k isometric embedding in Minkowski spacetime $E^{N,1}$, for some $N > 0$, if and only if (M, g) is stable.*

As for the optimal conditions for the validity of the distance formula with D replaced by d, i.e. that originally suggested by Parfionov and Zapatrin, we found that in Lorentzian manifolds endowed with C^1 metrics the formula holds if and only if the spacetime is causally continuous and the Lorentz-Finsler distance is finite and continuous [1]. In particular, globally hyperbolic spacetimes are of this type and in this case the family of functions can be further restricted, so for instance, the functions can be taken Cauchy.

3. A second proof

In this section we give a different proof of the smooth distance formula for Lorentzian globally hyperbolic spacetimes endowed with $C^{2,1}$ metrics. The proof might be adapted to the Lorentz-Finsler case and to weaker regularity assumptions, however there seems to be no point in pursuing this direction since the strategy of the previous section really allowed us to prove a stronger version under much weaker conditions.

The approach of this section was really the first one employed by the author to prove the distance formula. Though it has the limitation that it does not give the optimal conditions for the existence of smooth steep time functions, it is otherwise perfectly fine if one is interested in globally hyperbolic spacetimes endowed with sufficiently regular metrics.

The proof uses Theorem 2.2, a previous construction of steep functions by the author [17], stability results for Cauchy temporal functions and Lorentzian distance [1], and a previous non-differentiable version of the distance formula proved by Franco [12] (thus it is certainly not self contained). For shortness the final step in the proof assumes familiarity with Franco's paper and more generally with the details of [12,17]. A *Cauchy* time function is a time function whose restriction to any inextendible continuous causal curve has image $\mathbb{R}$. In the next theorems $\mathscr{F}$ is as in Eq. (1).

Theorem 3.1. *Let (M, g) be a globally hyperbolic spacetime and let h be any Riemannian metric on M. There is a smooth h-steep Cauchy temporal function τ. Moreover, there is a smooth h-steep Cauchy temporal function which is $\mathscr{F}$-steep.*

This theorem was also obtained by Suhr in a recent work using different methods [18, Th. 2.3].

Proof. The proof of the existence of a smooth Cauchy $\mathscr{F}$-steep time function in globally hyperbolic spacetimes can be found in [17] (for a previous proof, see [16]). One can prove more. Since global hyperbolicity is stable [5, 19] one can find a smooth Cauchy steep time function τ' for a Lorentzian Finsler function $\mathscr{F}'$ (cf. (1)) with larger cones $C' > C$, such that the indicatrix $\mathscr{F}'^{-1}(1)$ of $\mathscr{F}'$ does not intersect that of $\mathscr{F}$. So for every $v \in C_x$, $\mathrm{d}\tau'(v) \ge \mathscr{F}'(v)$ with the bonus that now $\mathscr{F}'$ does not vanish on lightlike vectors, i.e. on ∂C_x.

Let h be any auxiliary Riemannian metric. For every $x \in M$, at T_xM we can shrink through homotheties the indicatrix of $\mathscr{F}'$, so redefining this function but not its cone, in such a way that its indicatrix intersected with C_x is contained in $B^h(x, 1)$ (the open unit ball centered at x with respect to the distance induced by h). As a consequence, for the Cauchy $\mathscr{F}'$-steep time function τ' we have the inclusion $\{v : \mathrm{d}\tau'(v) = 1\} \cap C_x \subset B^h(x, 1)$, which implies $\mathrm{d}\tau'(v) \ge \mathscr{F}'(v) \ge \|v\|_h$ for every $v \in C$. But the fact that the indicatrix of $\mathscr{F}'$ does not intersect that of $\mathscr{F}$ also implies $\mathscr{F}'(v) \ge \mathscr{F}(v)$ for every $v \in C$. Thus with the redefinition $\tau' \to \tau$ we get the desired result. □

Theorem 3.2. *Let h be an auxiliary Riemannian metric. In a globally hyperbolic spacetime (M, g) both topology and order can be recovered from the set $\mathscr{V}$ of smooth Cauchy h-steep time functions. That is:*

IOP Conf. Series: Journal of Physics: Conf. Series **968** (2018) 012009 doi:10.1088/1742-6596/968/1/012009

(a) $(x, y) \in J \Leftrightarrow t(x) \le t(y)$, *for every* $t \in \mathscr{V}$;

(b) for every open set $O \ni p$ *we can find* $f, h \in \mathscr{V}$ *in such a way that* $p \in \{q\colon f(q) > 0\} \cap \{q\colon h(q) < 0\} \subset O$.

Remark 3.3. Thus topology and order can be recovered from the set of smooth Cauchy $\mathscr{F}$-steep temporal functions. In fact, if we take h so that its balls do not intersect the indicatrices of $\mathscr{F}$ we have that every smooth Cauchy h-steep time function is a smooth Cauchy $\mathscr{F}$-steep temporal function. Notice that h can be chosen complete.

Proof. Let $q \notin J^+(p)$ then the spacetime $N = M\backslash(J^+(p) \cup J^-(q))$ is globally hyperbolic. Any Cauchy hypersurface S for N is also a Cauchy hypersurface for M with $p \in I^+(S)$ and $q \in I^-(S)$. The Geroch topological splitting theorem [14] implies that we can find a Cauchy time function t for M so that $S = t^{-1}(0)$, $t(p) > 1$, $t(q) < -1$ (for instance, apply the theorem to $M\backslash J^-(S)$ and to $M\backslash J^+(S)$ and reparametrize the level sets so obtained into a global Cauchy time function). Let S_t be its constant slices. Inspection of the proof in [17] shows that we can find a smooth Cauchy $\mathscr{F}$-steep temporal function t' such that $S'_0 := t'^{-1}(0) \subset t^{-1}([-1, 1])$, $t' > t$ for $t > 1$, and $t' < t$ for $t < -1$. An improvement [1, Th. 47] over the classical stability result for global hyperbolicity [1, 5, 19, 21] states that the Cauchy temporal functions are stable so we can widen the cones while preserving global hyperbolicity and the Cauchy property of S'_0. Let $\mathscr{F}'$ be a Lorentzian Finsler function for a wider cone structure, $C' > C$, chosen so that the indicatrix $\mathscr{F}'^{-1}(1)$ intersected with C is contained in the unit ball of h, and hence so that any $\mathscr{F}'$-steep time function f satisfies $\mathrm{d}f(v) > \|v\|_h$ for $v \in C$. Repeating the above argument we find a smooth Cauchy $\mathscr{F}'$-steep temporal function t'' such that $S''_0 \subset t'^{-1}([-1, 1])$, $t'' > t'$ for $t' > 1$, and $t'' < t'$ for $t' < -1$. In particular $t''(p) > 1 > 0 > -1 > t''(q)$. This result proves the representability of the order by smooth h-steep Cauchy time functions.

For the topology, let $p \in O$, O open, and let t be a smooth Cauchy h-steep time function such that $t(p) = 0$, cf. Th. 3.1. Let $S_0 = t^{-1}(0)$ and let $Q \subset O$ be a compact neighborhood of p such that $Q = J^+(Q) \cap J^-(S_0) \cup J^-(Q) \cap J^+(S_0)$, and let φ be a positive smooth function supported on Q. Let τ^-_φ be the volume function [15]. We know that τ^-_φ is C^1 with past directed timelike gradient wherever $E^-(q)$ intersect the locus $\varphi > 0$ and with a vanishing differential otherwise. Thus $f := t + \tau^-_\varphi$ is a C^1 Cauchy h-steep time function. Notice that the locus $f = 0$ coincides with $t = 0$ outside Q, $f(p) > 0$ and $f > 0$ only for $t > 0$ or inside Q. Inverting the time orientation we get a C^1 Cauchy h-steep time function u such that $u = 0$ coincides with $t = 0$ outside Q, $u(p) < 0$ and $u < 0$ only for $t < 0$ or inside Q. Thus $p \in \{q\colon f(q) > 0\} \cap \{q\colon u(q) < 0\} \subset O$. The smooth version of this inclusion is due to the density of $C^\infty(M, \mathbb{R})$ in $C^1(M, \mathbb{R})$, see [20, Th. 2.6]. □

Theorem 3.4. *Let* (M, g) *be a globally hyperbolic spacetime and let* $\mathscr{U}$ *be the family of smooth, Cauchy,* $\mathscr{F}$*-steep temporal functions which are* h*-steep for some complete Riemannian metric* h *(dependent on the function). We have the identity*

$$d(p, q) = \inf\{[f(q) - f(p]^+\colon f \in \mathscr{U}\}. \tag{13}$$

Proof. Suppose first that $(p, q) \notin J$ so that $d(p, q) = 0$. We know from Remark 3.3 that the functions in $\mathscr{U}$ represent the order, so there is $f \in \mathscr{U}$ such that $f(p) > f(q)$, thus $d(p, q) = 0 = [f(q) - f(p)]^+$.

Now, suppose that $(p, q) \in J$. Let f be smooth and $\mathscr{F}$-steep. Given a continuous causal curve $x\colon [0, 1] \to M$ connecting p to q,

$$f(q) - f(p) = \int_x \dot{f}\mathrm{d}t = \int_x \mathrm{d}f(\dot{x})\mathrm{d}t \ge \int_x \mathscr{F}(\dot{x})\mathrm{d}t = \ell(x).$$

IOP Conf. Series: Journal of Physics: Conf. Series **968** (2018) 012009 doi:10.1088/1742-6596/968/1/012009

Thus taking the supremum over the connecting continuous causal curves we get $d(p,q) \le f(q) - f(p) = [f(q) - f(p)]^+$, and taking the infimum over the family of functions we obtain the inequality $\le$.

For the other inequality we have to show that for every $\epsilon > 0$ we can find a smooth Cauchy $\mathscr{F}$-steep temporal function f such that $f(q) - f(p) \le d(p,q) + \epsilon$. We can enlarge the cones while preserving global hyperbolicity. In particular, we can find a Lorentzian $\hat{\mathscr{F}}$ globally hyperbolic, with $\hat{C} > C$ such that the indicatrices inside the cones do not intersect and $d(p,q) \le \hat{d}(p,q) \le d(p,q) + \epsilon/4$. The first inequality is clear, while the second inequality follows from [1, Th. 58,59,61].

Let K be a compact neighborhood of $\hat{J}^+(p) \cap \hat{J}^-(q)$. Notice that apart from the definition of the cone $\hat{C}$, so far $\hat{\mathscr{F}}$ is unconstrained outside K. Let h' be a complete Riemannian metric. We choose the indicatrices $\hat{\mathscr{I}}$ of $\hat{\mathscr{F}}$ outside K to be so close to the origin that their intersection with C is contained in the unit sphere bundle of h', that is $\hat{\mathscr{F}}(v) > |v|_{h'}$ for every $v \in C$.

Notice that since the indicatrices $\hat{\mathscr{I}}$ and $\mathscr{I}$ do not intersect we can find a Riemannian metric h so small that for every $v \in C$, $\hat{\mathscr{F}}(v) \ge \mathscr{F}(v) + 2|v|_h$ and outside K the unit balls of h contain those of h' of radius 2, that is $|\cdot|_{h'} > 2|\cdot|_h$ outside K.

Let us apply Franco's construction to $(M, \hat{\mathscr{F}})$. Inspection of his proof shows that he constructs a continuous function $\hat{f}$ such that $\hat{f}(q) - \hat{f}(p) < \hat{d}(p,q) + \epsilon/4$ by means of a (locally finite) sum of functions of the form $\hat{d}^+_r(\cdot) := \hat{d}(r,\cdot)$, $\hat{d}^-_r(\cdot) := -\hat{d}(\cdot, r)$, where r runs over a countable set of points $\{r_i\}$. Any point of the manifold belongs to the support of one of these functions. As a consequence, for every $(a,b) \in \hat{J}$, Franco's function satisfies $\hat{f}(b) - \hat{f}(a) \ge \hat{\ell}(x)$ where x is any $\hat{C}$-causal curve connecting a and b (just partition the causal curve so that each part belongs to the support of a function $\hat{d}^+_r$ or $\hat{d}^-_r$ and use the reverse triangle inequality for $\hat{d}$).

We can use Theorem 2.2 with $\underline{F} = \hat{\mathscr{F}}$, thus we can find f smooth such that $|f - \hat{f}| < \epsilon/4$ and $\mathrm{d}f(v) \ge \hat{\mathscr{F}}(v) - |v|_h \ge \mathscr{F}(v) + |v|_h$ on every C-causal vector v. In particular, f is h-steep and $\mathscr{F}$-steep. But we have also $\hat{\mathscr{F}}(v) - |v|_h \ge |v|_{h'} - |v|_h \ge \frac{1}{2}|v|_{h'}$ outside K, thus f is $h'/4$-steep outside K and h-steep inside K, which implies that f is h''-steep for some complete Riemannian metric h'', and hence Cauchy. Finally,

$$\begin{aligned} f(q) - f(p) &\le \hat{f}(q) - \hat{f}(p) + |\hat{f}(q) - f(q)| + |\hat{f}(p) - f(p)| \le \hat{d}(p,q) + \tfrac{3}{4}\epsilon \\ &\le d(p,q) + \epsilon. \end{aligned}$$

□

References

[1] Minguzzi E 2017 Causality theory for closed cone structures with applications arXiv:1709.06494
[2] Nachbin L 1965 *Topology and order* (Princeton: D. Van Nostrand Company, Inc.)
[3] Minguzzi E 2013 *Topol. Appl.* **160** 965–978 arXiv:1212.3776
[4] Aubin J P and Cellina A 1984 *Differential inclusions* (*Grundlehren der Mathematischen Wissenschaften [Fundamental Principles of Mathematical Sciences]* vol 264) (Springer-Verlag, Berlin)
[5] Fathi A and Siconolfi A 2012 *Math. Proc. Camb. Phil. Soc.* **152** 303–339
[6] Minguzzi E 2009 *Commun. Math. Phys.* **290** 239–248 arXiv:0809.1214
[7] Minguzzi E 2010 *Commun. Math. Phys.* **298** 855–868 arXiv:0909.0890
[8] Bernard P and Suhr S Lyapounov functions of closed cone fields: from Conley theory to time functions arXiv:1512.08410v2. Replaces a previous work by Suhr "On the existence of steep temporal functions".
[9] Connes A 1994 *Noncommutative geometry* (Academic Press, Inc., San Diego, CA)
[10] Parfionov G N and Zapatrin R R 2000 *J. Math. Phys.* **41** 7122–7128
[11] Moretti V 2003 *Rev. Math. Phys.* **15** 1171–1217
[12] Franco N 2010 *SIGMA Symmetry Integrability Geom. Methods Appl.* **6** Paper 064, 11
[13] Hawking S W 1968 *Proc. Roy. Soc. London, series A* **308** 433–435
[14] Hawking S W and Ellis G F R 1973 *The Large Scale Structure of Space-Time* (Cambridge: Cambridge University Press)

IOP Conf. Series: Journal of Physics: Conf. Series **968** (2018) 012009 doi:10.1088/1742-6596/968/1/012009

[15] Chruściel P T, Grant J D E and Minguzzi E 2016 *Ann. Henri Poincaré* **17** 2801–2824 arXiv:1301.2909
[16] Müller O and Sánchez M 2011 *Trans. Am. Math. Soc.* **363** 5367–5379
[17] Minguzzi E 2016 *Class. Quantum Grav.* **33** 115001 arXiv:1601.05932
[18] Suhr S On the existence of steep temporal functions arXiv:1512.08410
[19] Benavides Navarro J J and Minguzzi E 2011 *J. Math. Phys.* **52** 112504 arXiv:1108.5120
[20] Hirsch M W 1976 *Differential topology* (New York: Springer-Verlag)
[21] Sämann C 2016 *Ann. Henri Poincaré* **17** 1429–1455

IOP Conf. Series: Journal of Physics: Conf. Series **968** (2018) 012010 doi:10.1088/1742-6596/968/1/012010

On geodesics in low regularity

Clemens Sämann and Roland Steinbauer
Faculty of Mathematics, University of Vienna, Austria
E-mail: clemens.saemann@univie.ac.at, roland.steinbauer@univie.ac.at

Abstract. We consider geodesics in both Riemannian and Lorentzian manifolds with metrics of low regularity. We discuss existence of extremal curves for continuous metrics and present several old and new examples that highlight their subtle interrelation with solutions of the geodesic equations. Then we turn to the initial value problem for geodesics for locally Lipschitz continuous metrics and generalize recent results on existence, regularity and uniqueness of solutions in the sense of Filippov.

Keywords: geodesics, extremal curves, low regularity, Filippov solutions

MSC2010: 53B30, 53C22, 34A36, 83C99

1. Introduction

Recently there has been an increased interest in several low regularity aspects of general relativity and of semi-Riemannian geometry. In particular, it has been confirmed that the regularity class of $C^{1,1}$ (the metric possessing locally Lipschitz continuous first derivatives) can rightly be seen as the threshold of classical theory. In addition to the well-known fact that the geodesic equation can be locally uniquely solved by classical ODE-theory, the exponential map retains maximal regularity and convex neighborhoods exist [23, 17]. In the Lorentzian case the bulk of causality theory remains valid [4, 23, 18] as well as the classical singularity theorems [19, 20, 9]. On the other hand studies of geometries of regularity below $C^{1,1}$ (see e.g. [4, 32, 33, 8]) have revealed important differences in many of the well known concepts and have emphasized the significance of regularity issues.

In this contribution we focus on geodesics in low regularity in Riemannian as well as in Lorentzian manifolds. This topic again is of particular interest since many of the facts well known for smooth metrics fail to extend to a regularity below $C^{1,1}$. For example in the classical setting the unique local solutions of the geodesic equation are locally extremal curves for the length functional. Conversely extremal curves are pregeodesics. Moreover Lorentzian geodesics and maximal curves have a causal character.

In case of insufficient regularity the question of existence of extremal curves becomes a separate issue. It has long been answered affirmatively in the case of continuous Riemannian metrics and we will discuss the recent equally positive answer in the Lorentzian case. It might be surprising that already in cases where the geodesic equation is (classically locally) solvable but not uniquely so (in particular, for metrics of Hölder class $C^{1,\alpha}$ for any $\alpha < 1$) the connection between its solutions and extremal curves becomes subtle. We will discuss some classical and some new examples in that realm at some length. We will supplement them by examples (again old and new) which demonstrate the failure of some of the usual causality properties in Lorentzian manifolds.

Complementing this line of investigation we will transfer some recent results on solutions of the geodesic equation for impulsive gravitational wave spacetimes into a more abstract setting. In particular, we will discuss existence and regularity of solutions to the geodesic equation for locally Lipschitz semi-Riemannian metrics using the solution concept of Filippov [7] for ODEs with discontinuous right hand side. We also present some sufficient conditions for uniqueness, which have proved to be useful in the context of applications. Finally we provide an outlook to open questions and related problems.

We end this introduction by fixing some notions and notations. All manifolds will be assumed to be of class C^∞ (which is no loss of generality, [15, Thm. 2.9]) and we will only lower the regularity of the metric. Hence e.g. by a continuous spacetime (M, g) we will mean a smooth, connected manifold M of dimension $n \geq 2$ equipped with a continuous Lorentzian metric g with a time orientation induced by a (continuous) timelike vector field. Our notations are quite standard and we generally follow [25]. Mindful of the above discussion a geodesic will always mean a solution (of some sort) of the geodesic equation and should be strictly distinguished from extremal curves.

2. Extremal curves

As indicated in the introduction, below a regularity of $C^{1,1}$ the notions of extremal curves and geodesics no longer coincide. We discuss existence of extremal curves for continuous metrics and study their relation to geodesics. We start with the case of Riemannian metrics.

2.1. The Riemannian case

It is a classical result by Hilbert [14] (using the Theorem of Arzela-Ascoli) that for *continuous* Riemannian metrics minimizing curves always exist locally. The global existence of minimizing curves is a corollary of the Hopf-Rinow-Cohn-Vossen Theorem for length spaces [2, Thm. 2.5.28]. To be precise, by [3], a continuous Riemannian metric gives rise to a length space, and hence if the corresponding metric space is complete any two points can be connected via a minimizing curve.

So the situation of existence of minimizing curves for continuous Riemannian metrics is exactly as in the smooth case. However, the relation between such minimizing curves and geodesics turns out to differ drastically from the smooth situation if the regularity of the metric drops below $C^{1,1}$.

Indeed classical examples show the explicit failure of the initial value problem for geodesics to be *uniquely* solvable for $g \in C^{1,\alpha}$ for any fixed $\alpha < 1$ [12], while it is possible that at the same time all the usual properties hold [13]: minimizing curves are locally unique, the boundary value problem for geodesics is locally uniquely solvable and even in 'singular points' there is a locally minimizing curve starting off in any direction.

However, another classical example [13] shows that for $g \in C^{1,\alpha}$ again for any fixed $\alpha < 1$ geodesics need not be even minimizing locally. This example also shows that again even locally the boundary value problem for geodesics is non-uniquely solvable. As the same phenomenon

IOP Conf. Series: Journal of Physics: Conf. Series **968** (2018) 012010 doi:10.1088/1742-6596/968/1/012010

occurs also in the Lorentzian case by a modification of the original example we will discuss it in some detail below.

On the positive side the regularity of minimizing curves is slightly better than one would expect. A classical result in this direction is the following: If the metric is C^1, then minimizing curves are actually geodesics and are of regularity C^2. This can be seen by using the trick of Du Bois-Reymond [1, Ch. I, §6], which we briefly recall here.

Let g be a C^1-Riemannian metric. Since it suffices to argue locally we consider $\gamma\colon [a,b] \to \mathbb{R}^n$, a locally Lipschitz continuous and minimizing curve from $p = \gamma(a)$ to $q = \gamma(b)$. We parametrize γ by g-arclength, i.e., $g_{\gamma(t)}(\dot\gamma(t), \dot\gamma(t)) = 1$ almost everywhere. That γ is minimizing implies that

$$\frac{d}{d\varepsilon}\bigg|_{\varepsilon=0} L_g(\gamma + \varepsilon\phi) = 0\,,$$

where L_g is the length functional of g, i.e., $L_g(\gamma) = \int_a^b F(t,\gamma(t),\dot\gamma(t))\,\mathrm{d}t$ with $F(t,\gamma(t),\dot\gamma(t))= \sqrt{g_{\gamma(t)}(\dot\gamma(t),\dot\gamma(t))}$, and moreover ϕ is a smooth function with compact support in $[a,b]$. For simplicity we calculate the following in one dimension, however the general case poses no additional difficulties. Integration by parts yields

$$0 = \frac{d}{d\varepsilon}\bigg|_{\varepsilon=0} L_g(\gamma + \varepsilon\phi) = \int_a^b F_\gamma \phi + F_{\dot\gamma}\phi'\,\mathrm{d}t = \int_a^b (F_{\dot\gamma} - \int_a^t F_\gamma\,\mathrm{d}s)\phi'\,\mathrm{d}t\,. \tag{1}$$

Thus, since ϕ is arbitrary, $F_{\dot\gamma} - \int_a^t F_\gamma\,\mathrm{d}s$ is constant and in the general higher dimensional case this yields that

$$F_{\dot\gamma^i} - \int_a^t F_{\gamma^i}\,\mathrm{d}s = c_i = \text{constant}\,, \tag{2}$$

for $i = 1,\dots,n$. Note that γ minimizes on any subinterval $[a,s]$ $(s \le b)$, and hence we obtain from (2) in coordinates (x^i) $(\gamma^i = x^i \circ \gamma)$ that

$$g_{ij}\,\dot\gamma^j - \frac{1}{2}\int_a^s \frac{\partial g_{lj}}{\partial x^i}\dot\gamma^l\dot\gamma^j\,\mathrm{d}r = c_i\,,$$

and by multiplying with $g^{-1} = (g^{ij})$ this gives

$$\dot\gamma^m = g^{im}\left(\frac{1}{2}\int_a^s \frac{\partial g_{lj}}{\partial x^i}\dot\gamma^l\dot\gamma^j\,\mathrm{d}r + c_i\right)\,. \tag{3}$$

The right-hand-side of (3) is continuous because $\frac{\partial g_{lj}}{\partial x^i}$ is continuous and $\dot\gamma^i$ is bounded. Thus γ is C^1 and using this information one sees again by (3) that γ is in fact C^2.

Finally it follows that γ actually is a geodesic by the usual argument using integration by parts on the other term in (1).

A recent and enhanced regularity result is given in [21]: If the Riemannian metric is of Hölder-regularity $C^{0,\alpha}$ with $0 < \alpha \le 1$, then minimizing curves are of regularity $C^{1,\beta}$, where $\beta = \frac{\alpha}{2-\alpha}$. This regularity $C^{1,\beta}$ is optimal as there are examples of minimizers with respect to a $C^{0,\alpha}$ Riemannian metric on $\mathbb{R}^2$ that are not $C^{1,l}$ for $l > \beta$ [21, Thm. 1.1]

Finally, we review a key example of Hartman and Wintner given in [13].

Example 2.1. (The Hartman-Wintner example [13, Sec. 5]) We consider $M := (-1,1) \times \mathbb{R}$ with the metric

$$g_{(x,y)} = dx^2 + (1 - |x|^\lambda)dy^2\,, \tag{4}$$

where $1 < \lambda < 2$. The metric is of Hölder-regularity $C^{1,\lambda-1}$ and smooth off the y-axis. The geodesic equations are

$$x'' + \frac{\lambda}{2}|x|^{\lambda-1}\operatorname{sgn} x (y')^2 = 0, \qquad y'' - \frac{\lambda}{2}\frac{|x|^{\lambda-1}\operatorname{sgn}(x)}{1-|x|^\lambda}x'y' = 0\,. \tag{5}$$

Moreover, it suffices to consider only the initial value $(x_0, y_0) = (0,0)$ since the metric does not depend on y at all. Note that since the metric is C^1, $g_{\gamma(s)}(\dot\gamma(s), \dot\gamma(s))$ is constant and we parametrize any geodesic by arclength. The y-equation in (5) is equivalent to $(1-|x|^\lambda)y' = c =$ constant. In case $c = 0$ we have $y = 0$, so the x-equation in (5) is trivial and the geodesic is simply given by $y = 0$. If $c \neq 0$, then $y' \neq 0$ and y is strictly monotonous along any such geodesic. Parametrizing by arclength gives

$$x' = \pm\sqrt{1 - \frac{c^2}{1-|x|^\lambda}}\,, \tag{6}$$

with $c \in [-1,1]$. If $c^2 = 1$, then $x = 0$ is the only solution to (6) and we denote this geodesic by γ_0. If $c^2 < 1$, then the right-hand-side of (6) is (initially) C^1 and thus there is a unique solution with initial condition $x(0) = 0$. Given this unique solution x, we can uniquely solve the y-equation by integrating, i.e., $y(s) = \int_0^s \frac{c}{1-|x(r)|^\lambda}dr$. To summarize, this means that at $(0,0)$ for every initial direction there is a unique geodesic, and by symmetry the initial-value-problem is uniquely solvable for arbitrary data.

To determine the shape of the geodesics set $c = \pm\sqrt{1-\varepsilon}$ for $\varepsilon \in [0,1]$ and denote by $\gamma_{\pm\varepsilon}$ the geodesics starting at $(0,0)$ with initial velocity $(\pm\sqrt{\varepsilon}, \sqrt{1-\varepsilon})$, see Figure 1. A short calculation shows that γ_ε reaches $x = \sqrt[\lambda]{\varepsilon}$ in finite time. We denote the corresponding parameter value by s_0 and the y-value by y_1, i.e., $\gamma_\varepsilon(s_0) = (\sqrt[\lambda]{\varepsilon}, y_1)$. Then $\dot\gamma_\varepsilon(s_0) = (0, 1/\sqrt{1-\varepsilon})$ is vertical and since g is independent of y we can reflect γ_ε at $y = y_1$. Thus by symmetry $(0,0)$ is connected to $(0, 2y_1)$ by three distinct geodesics: $\gamma_{\pm\varepsilon}$ and γ_0. At this point we vary ε and it is not hard to see that $y_1(\varepsilon) \to 0$ as $\varepsilon \searrow 0$. This implies that the boundary value problem for geodesics is not uniquely solvable in any neighborhood of any point on the y-axis.

Finally, it is again not hard to see that the geodesic γ_0 is not minimizing between any of its points and that in fact $\gamma_{\pm\varepsilon}$ are minimizing between its endpoints. From here we obtain that even locally minimizing curves are not unique and there is no minimizing curve with initial velocity $(0,1)$.

2.2. The Lorentzian case

In complete analogy with the Riemannian case local existence of maximizing causal curves holds for *continuous* Lorentzian metrics. The key notion here is *global hyperbolicity*, which for continuous metrics has been studied in [32]. There a spacetime is called globally hyperbolic if it is non-totally imprisoning and the causal diamonds are compact. Recall that a *Cauchy hypersurface* is a set $S \subseteq M$ that is met exactly once by any inextendible (locally Lipschitz) causal curve. The existence of a Cauchy hypersurface is equivalent to global hyperbolicity

IOP Conf. Series: Journal of Physics: Conf. Series **968** (2018) 012010 doi:10.1088/1742-6596/968/1/012010

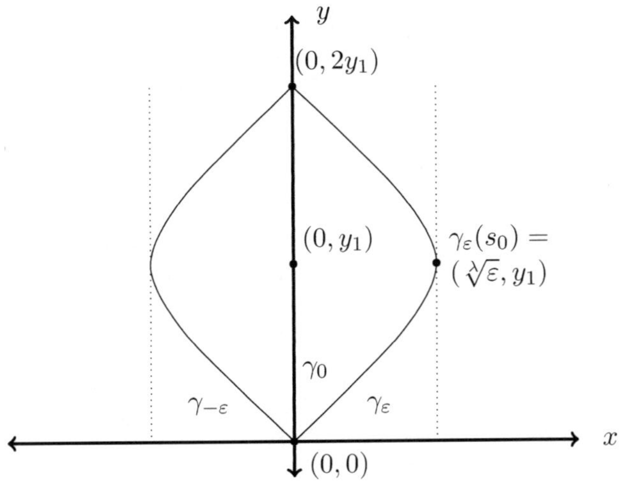

Figure 1: The minimizing geodesics $\gamma_{\pm\varepsilon}$ and the non-minimizing one γ_0 for the metric (4).

also in this low regularity [32, Thm. 5.7 and Thm. 5.9]. We may establish without much effort:

Theorem 2.2. (Existence of locally maximizing curves) Every point p in a continuous spacetime (M, g) possesses a neighborhood U such that any two U-causally related points can be joined by a maximizing (in U) causal curve.

Proof: Let $\hat{g}$ be a smooth Lorentzian metric on M, with the property that the lightcones of g are contained in the timelike cones of $\hat{g}$ (see [4]). Let $p \in M$ then there exists a $\hat{g}$-globally hyperbolic neighborhood (U, x^μ) of p by [24, Thm. 2.14], i.e., in the x^μ-coordinates one has that $x^0 = 0$ is a Cauchy hypersurface in U with respect to $\hat{g}$. Then obviously this is also a Cauchy hypersurface with respect to g. Hence $(U, g|_U)$ is globally hyperbolic by [32, Thm. 5.7] and thus maximal (in U) causal curves exist between any two (in U) causally related points by [32, Prop. 6.4]. □

Here we have already made use of the Avez-Seifert result for continuous metrics which we quote below as the global analogue of the above result. Note, however, that Theorem 2.2 can be established without using the global result: one may just use that there are local bounds on the arclength of causal curves with respect to a complete Riemannian metric and then use the Theorem of Arzela-Ascoli in the form of the limit curve theorem [22, 4, 32].

Theorem 2.3. (Avez-Seifert, [32, Prop. 6.4]) Let (M, g) be a globally hyperbolic continuous spacetime. Then there is a maximal causal curve connecting any two points in the spacetime which are causally related.

We summarize the Riemannian and Lorentzian results on the existence of minimal and maximal curves in the following table.

IOP Conf. Series: Journal of Physics: Conf. Series **968** (2018) 012010 doi:10.1088/1742-6596/968/1/012010

$g \in C^0$	local	global
Riemannian	length sructure, Arzela-Ascoli [14]	Hopf-Rinow-Cohn-Vossen in complete length spaces [2, Thm. 2.5.28], [3]
Lorentzian	existence of globally hyperoblic neighborhoods	continuous Avez-Seifert [32]

Again as in the Riemannian case the relation of maximizing causal curves with geodesics becomes subtle for a regularity below $g \in C^{1,1}$. Indeed Example 2.1 can be modified to yield a Lorentzian counterexample.

Example 2.4. (The Lorentzian Hartman-Wintner example) We consider $M := \mathbb{R}\times(-1,1)\times\mathbb{R}$ with the $C^{1,\lambda-1}$-metric

$$g_{(t,x,y)} = -dt^2 + dx^2 + (1-|x|^\lambda)dy^2\,,$$

where again $1<\lambda<2$. Then the geodesic equations for x and y are just equations (5) and in addition we have the trivial equation $t''=0$.

Again it suffices to consider geodesics starting at the origin. By Example 2.1 the initial value problem is uniquely solvable and we again consider some special solutions. Defining y_1 as in Example 2.1 we consider the timelike geodesic $\Gamma_0(r) = (2\sqrt{2}s_0 r, 0, 2y_1 r)$ $(r \in [0,1])$. To see that Γ_0 is timelike, note that $s_0 < y_1 < 2s_0$. Moreover $L(\Gamma_0) = \sqrt{8s_0^2 - 4y_1^2} < 2s_0$. On the other hand set $\Gamma_{\pm\varepsilon}(s) = (\sqrt{2}s, \gamma_{\pm\varepsilon}(s))$ $(s \in [0, 2s_0])$, where $\gamma_{\pm\varepsilon}$ are the minimizing geodesics of Example 2.1. Then the geodesics $\Gamma_{\pm\varepsilon}$ are normalized to eigentime and so the lengths $L(\Gamma_{\pm\varepsilon}) = 2s_0$ coincide with parameter length. Therefore we obtain without effort similar results as in the original example.

Indeed $\Gamma_{\pm\varepsilon}(s_0) = (\sqrt{2}s_0, \pm\sqrt[\lambda]{\varepsilon}, y_1)$. Then the points $(0,0,0)$ and $(2\sqrt{2}s_0, 0, 2y_1)$ are connected by the three distinct geodesics $\Gamma_{\pm\varepsilon}$ and Γ_0 and again the boundary value problem for geodesics is not uniquely solvable. Moreover, Γ_0 is not maximizing between any of its points since the timelike curves $\Gamma_{\pm\varepsilon}$ are longer: $L(\Gamma_0) < 2s_0 = L(\Gamma_{\pm\varepsilon})$. Finally there is no maximizing curve with initial velocity$(\sqrt{2}s_0, 0, y_1)$ and by symmetry maximizing curves are not unique.

This example is complemented by the following one from [4] which, in particular, shows that the push-up principle of causality theory fails to hold in Hölder-regularity $C^{0,\lambda}$ with $\lambda < 1$. Recall that the push-up principle states that if two points can be connected via a causal curve that contains a timelike segment, then the points can be connected by a timelike curve. This Hölder-regularity is optimal as the push-up principle holds for (locally) Lipschitz continuous metrics [4, Cor. 1.17 and Prop. 1.23].

Example 2.5. (Bubbling [4, Ex. 1.11]) We consider the manifold $\mathbb{R}^2$ with the metric

$$g_{(u,x)} = -du^2 + 2(|u|^\lambda - 1)du\,dx + |u|^\lambda(2-|u|^\lambda)dx^2\,, \qquad (7)$$

where $0<\lambda<1$ and ∂_u gives the time-orientation. The metric is λ-Hölder regular and smooth of the x-axis. Null curves branch off from the x-axis and points in the region between the first branching null curve and the x-axis (which is also null) can be connected to the origin via causal curves (even of positive length) but not by timelike curves, see [4, Fig. 1.1]. This region is called the *causal bubble* and this demonstrates the failure of the push-up principle.

A slight modification of the above example yields the existence of a maximal causal curve in a $C^{0,\lambda}$-spacetime which does not have a causal character. This example can be found in detail in [16, Cor. 5.5].

Example 2.6. In the above Example 2.5 we put for simplicity $\lambda = \frac{1}{2}$ and consider only the manifold $M := (-1,1) \times \mathbb{R}$. Let $q := (u_0, x_0)$ be in the (upper right) *bubble region*, i.e., $0 < u_0 < \min(\frac{x_0}{4}, 1)$. As this spacetime is strongly causal and the causal diamond of the origin and q is compact, there is a maximal causal curve from the origin to q. This maximal causal curve cannot have a causal character: The curve has to first go along the x-axis, where it is null and after it leaves the x-axis it is in the smooth part of the spacetime. Moreover, there it has to have positive length and thus must be timelike.

Finally, we remark that the Du Bois-Reymond trick does not work in the Lorentzian setting. If g is a C^1-Lorentzian metric and γ is a maximal causal curve, then it is not clear that γ has to be timelike even if γ has positive length. Therefore, one cannot ensure that the variation $\gamma + \varepsilon\phi$ (compare the discussion the Riemannian case in Section 2.1) is causal, hence there is no reason that $\frac{d}{d\varepsilon}\big|_{\varepsilon=0} L_g(\gamma + \varepsilon\phi) = 0$. The argument only works out in case we already know that the maximizing curve is timelike.

3. The geodesic equation for locally Lipschitz metrics

In this section we deal with the geodesic equation for semi-Riemannian manifolds (M, g) with $g \in C^{0,1}$. The motivation for studying this regularity class comes on the one hand from the fact that here some crucial aspects of causality theory remain valid, cf. [4] and Section 2.2 above, and on the other hand that relevant exact solutions of general relativity such as impulsive gravitational wave spacetimes have this regularity.

Recall that in this regularity class the right-hand-side of the geodesic equation is only guaranteed to be locally bounded but *not* to be continuous. It turns out that a fruitful way to deal with these 'ODEs with discontinuous right hand side' is to use the Filippov solution concept [7], which we will now briefly review (see [5] for an application-driven introduction).

3.1. Filippov solutions

The key idea of Filippov's approach is to replace an ODE

$$\dot{x}(t) = f(x(t)) \qquad (t \in I)\,, \tag{8}$$

(I is some interval, $f\colon \mathbb{R}^n \supseteq D \to \mathbb{R}^n$ possibly of low regularity) by a *differential inclusion*

$$\dot{x}(t) \in \mathcal{F}[f](x(t)), \tag{9}$$

where the *Filippov set-valued map* $\mathcal{F}[f]\colon D \to \mathcal{K}_0(\mathbb{R}^n)$ (the collection of all nonempty, closed and convex subsets of $\mathbb{R}^n$) associated with F is defined as its essential convex hull

$$\mathcal{F}[f](x) := \bigcap_{\delta>0} \bigcap_{\mu(S)=0} \mathrm{co}\Big(f(B(x,\delta) \setminus S)\Big). \tag{10}$$

Here co denotes the closed convex hull, $B(x,\delta)$ is the closed ball of radius δ centered at x, and μ is the Lebesgue measure, cf. [31, Sec. 2].

The key idea is to look at the set of values f takes near a point of discontinuity. This fact lies also at the heart of the compatibility of this approach with approaches based on

IOP Conf. Series: Journal of Physics: Conf. Series **968** (2018) 012010 doi:10.1088/1742-6596/968/1/012010

regularization, see [11, Sec. 3.3]. The explicit calculation of a Filippov set-valued map can be non-trivial, however, there exists a calculus to at least bound this set [26]. Also note that a Filippov set-valued map is actually multi-valued only at the points of discontinuity of f.

Now a *Filippov solution* of the ODE (8) is defined to be an absolutely continuous curve $x\colon J \to U$, defined on some interval $J \subseteq I$, that satisfies the inclusion relation (9) almost everywhere.

Recall that a curve $x : [a,b] \to \mathbb{R}^d$ is said to be *absolutely continuous* if for every $\varepsilon > 0$ there is a $\delta > 0$ such that for all collections of non-overlapping intervals $([a_i, b_i])_{i=1}^m$ in $[a,b]$ with $\sum_{i=1}^m (b_i - a_i) < \delta$ we have that $\sum_{i=1}^m \|x(b_i) - x(a_i)\| < \varepsilon$. Moreover, recall that an absolutely continuous curve is differentiable almost everywhere.

Of course, if f is continuous, classical $\mathcal{C}^1$-solution and Filippov solutions coincide, but the latter exist under much more general assumptions on f. In fact, Filippov in [7] has developed a complete theory of ordinary differential equations based on this solution concept which has been found to be widely applicable e.g. in non-smooth mechanics. Here we just state the main existence result.

Theorem 3.1. ([7, Thm. 7.1]) The initial value problem

$$\dot{x}(s) \in A(s, x(s)) \text{ a.e.}, \quad x(t_0) = x_0 \qquad (t_0, x_0) \in I \times \mathbb{R}^n \tag{11}$$

has an absolutely continuous solution if the set valued map $I \times \mathbb{R}^n \ni (t,x) \mapsto A(t,x)$ satisfies

(i) $t \mapsto A(t,x)$ is Lebesgue measurable on I for all fixed x,

(ii) $x \mapsto A(t,x)$ is upper semi-continuous for almost all t, and

(iii) $\sup_{x\in\mathbb{R}^n} |A(t,x)| \le \beta(t) \in L^1_{\text{loc}}(I)$ for almost all t.

Our main interest lies in the following simple consequence, where we denote by L^∞_{loc} the space of all measurable and locally essentially bounded functions.

Corollary 3.2. ([7, Thm. 7.8]) If $f \in L^\infty_{\text{loc}}(D, \mathbb{R}^n)$, then for each $(t_0, x_0) \in I \times D$ there is a Filippov solution x of (8) with $x(t_0) = x_0$.

3.2. Existence of geodesics

This results lends itself to an application to the geodesic equations on semi-Riemannian manifolds M with a $C^{0,1}$-metric. First observe that the notion of the essential convex hull is invariantly defined on M. Indeed $\mathcal{F}(f)(\phi(x)) = \mathcal{F}(f \circ \phi)(x)$ for any diffeomorphism ϕ since the respective balls in (10) $B(\phi(x), r)$ and $\phi(B(x, r'))$ can be nested. Now rewriting the geodesic equations as a first order system one locally obtains from the Lipschitz property of the metric an equation of the form (8) with $f \in L^\infty_{\text{loc}}$. Hence we have:

Corollary 3.3. ([34, Thm. 2]) Let (M, g) be a smooth manifold with a $\mathcal{C}^{0,1}$-semi-Riemannian metric. Then there exist a Filippov solution of the geodesic equation for arbitrary data $p \in M$, $v \in T_pM$. These solutions possess absolutely continuous velocities (hence, in particular, are $\mathcal{C}^1$-curves).

Observe that such geodesics do not satisfy the geodesic equation in the classical sense. We only know that for almost all values of an affine parameter we have that $\ddot{\gamma}^i \in \mathcal{F}(-\Gamma^i{}_{jk}\dot{\gamma}^j\dot{\gamma}^k)$. Consequently, in general the norm of the tangent $|\dot{\gamma}|$ will not be preserved and hence in the Lorentzian case γ will not have a global causal character. For the same reason we do not know whether these geodesics are locally minimizing resp. maximizing curves. Also concerning regularity we do not know whether the geodesics are $C^{1,1}$, which would match the limiting case $\alpha = 1 = \beta$ in [21], see Section 2.1 above. In fact, we only obtain that the geodesics are C^1 with absolutely continuous tangent.

3.3. Uniqueness of geodesics

Next we turn to uniqueness. Clearly we cannot hope for any general result and we will rather be interested in situations where the $C^{0,1}$-metric is actually smooth off some hypersurface. This allows us to cover the case of 'matched spacetimes' and impulsive gravitational wave geometries, see below. Here we start with a general discussion of uniqueness of Filippov solutions for (8). In principle essentially one-sided Lipschitz conditions, i.e., $(f(x)-f(y))^t(x-y) \leq L\|x-y\|_2^2$ for some L and almost all x, y, provide one-sided uniqueness results (e.g. [5, Prop. 4]) without requiring f to be continuous. However, these results are ill-suited in case f is smooth off some hypersurface in $\mathbb{R}^n$ since piecewise smooth functions generically fail to be essentially one-sided Lipschitz even in cases where the corresponding Filippov solutions are unique, cf. [5, p. 53].

Hence we follow a different route and start by making the situation precise. Suppose that $f \in L^\infty_{\text{loc}}(D)$ and that the domain $D \subseteq \mathbb{R}^n$ is connected and a disjoint union $D = D^- \dot\cup N \dot\cup D^+$, where D^+, D^- are open sets and their common boundary, $N = \partial D^+ = \partial D^-$ is a smooth hypersurface, see the figures below. Now suppose $f \in C^1(D^\pm)$ up to the boundary N and denote by $f^\pm$ the extensions of $f|_{D^\pm}$ to the closure $\bar{D}^\pm = D^\pm \cup N$. Finally write $f_N^\pm$ for the projections of $f^\pm|_N$ on the unit normal $\vec{n}$ of N, pointing from D^- to D^+. We will abbreviate this situation by saying that the locally bounded function f is *smooth off the hypersurface* N. Then basic ODE-theory on the domains $\bar{D}^\pm$ yields the following:

Lemma 3.4. (Simple conditions for uniqueness, [7, Lem. 10.2]) Let $f \in L^\infty_{\text{loc}}(D, \mathbb{R}^n)$ be smooth off the hypersurface[1] N. If for $x_0 \in N$ we have $f_N^+(x_0) > 0$, then in $\bar{D}^+$ there exists a unique C^2-solution of (8) starting at x_0. Analogous assertions hold for $\bar{D}^-$ and $f_N^-(x_0) < 0$.

This basic observation also leads to the following conclusions: If at a point $x_0 \in N$ we have that f^+ points into D^+ ($f_N^+ > 0$) and f^- points into D^- ($f_N^- < 0$) then there are C^2-solutions which proceed into D^+ and such that proceed into D^- and one speaks of *repulsive trajectories*, see Figure 2. Clearly in this case uniqueness of Filippov solutions on D fails.

In all other possible cases uniqueness of Filippov solutions can be secured ([5, Prop. 5]). However, if at $x_0 \in N$ we have that f^+ points into D^- ($f_N^+ < 0$) and f^- points into D^+ ($f_N^- > 0$) then the C^2-solutions from either side may be trapped in N, a situation referred to as *sliding motion*, see Figure 3. We will be mainly interested in the remaining two cases, i.e., when the Filippov solutions cross from D^- into D^+ or vice versa. Indeed in the cases where f_N^+ and f_N^- share their sign, i.e., $f_N^\pm > 0$ (Figure 4) and $f_N^\pm < 0$ (Figure 5) solutions pass from D^- into D^+ and from D^+ to D^-, respectively. In this case one speaks of *transversally crossing trajectories*. More precisely the following criterion holds:

[1] In the sense defined above.

IOP Conf. Series: Journal of Physics: Conf. Series **968** (2018) 012010 doi:10.1088/1742-6596/968/1/012010

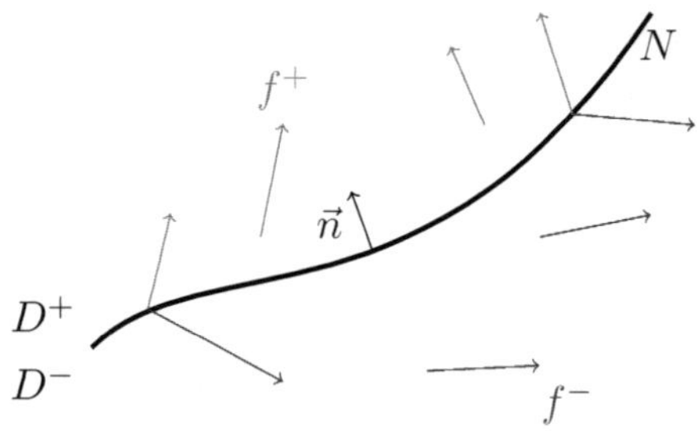

Figure 2: repulsive trajectories

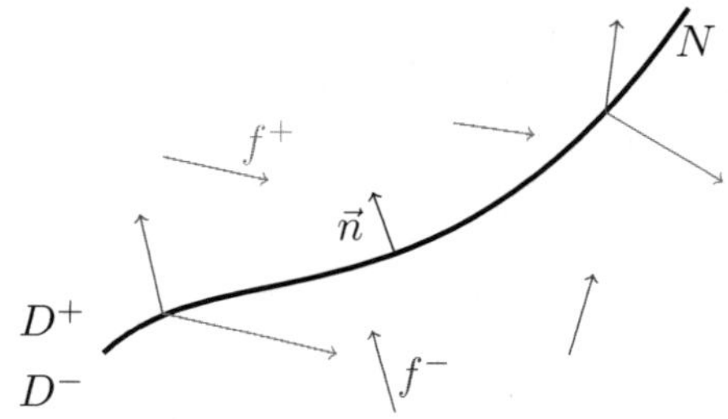

Figure 3: sliding motion

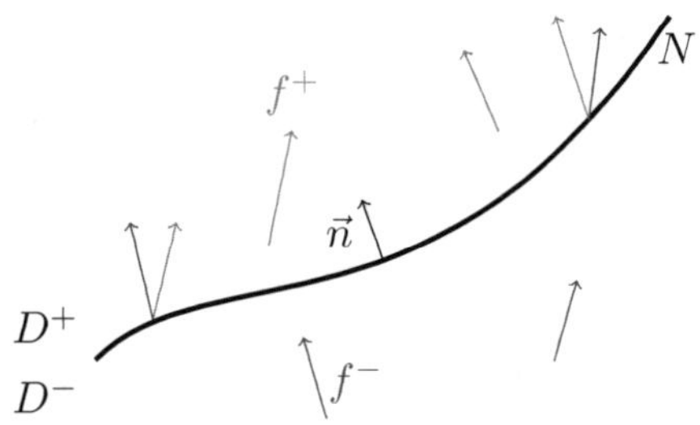

Figure 4: (upward) transversally crossing

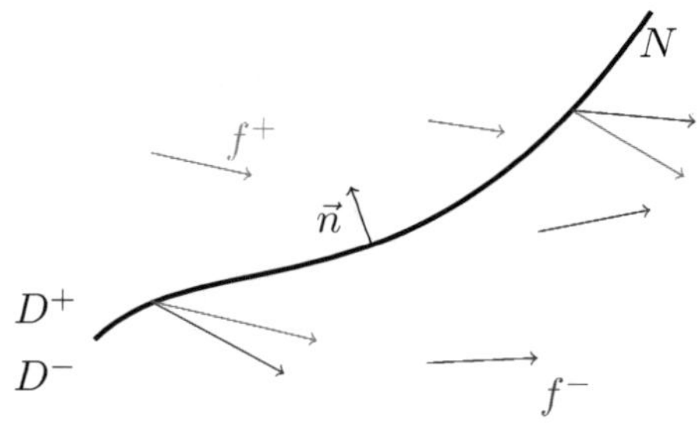

Figure 5: (downward) transversally crossing

Corollary 3.5. (Sufficient conditions for uniqueness [7, Cor. 10.1]) Let f be as above. On the region of the surface N where $f_N^+ > 0$ and $f_N^- > 0$, Filippov solutions that reach N from D^- pass to D^+ and hence uniqueness is not violated. The analogous assertion holds for $f_N^+, f_N^- < 0$ and solutions passing from D^+ to D^-.

We will now apply Corollary 3.5 to a geometric scenario that generically arises e.g. in 'matched spacetimes'. Essentially the present discussion isolates the abstract core of the one given in [27, Sec. 3.3] by neglecting the specific form of the spacetime metric used there. We start with a semi-Riemannian manifold (M, g) with $g \in C^{0,1}$ and assume that g is *smooth off a* C^∞*-hypersurface* $\mathcal{N}$ in the following sense: M is the disjoint union of some open sets $\mathcal{D}^+$, $\mathcal{D}^-$ and their common boundary $\mathcal{N} = \partial\mathcal{D}^+ = \partial\mathcal{D}^-$ which we assume to be a smooth hypersurface of M. The metric g is smooth in $\mathcal{D}^\pm$ up to the boundary, that is $g \in C^2(\mathcal{D}^\pm \cup \mathcal{N})$. (Here $\mathcal{D}^\pm \cup \mathcal{N}$ are smooth manifolds with boundary.)

Now a (Filippov solution of the) geodesic (equation) in $\mathcal{D}^\pm$ is a classical geodesic. Let us consider such a curve starting, say in D^- (the case D^+ being analogous) and reaching $\mathcal{N}$. Locally in coordinates $(\mathcal{U}, (x^1, \ldots, x^n))$ we write $\mathcal{N}$ as $\{x^1 = 0\}$ and we rewrite the geodesic equation for $\gamma(t) = (x^1(t), \ldots, x^n(t))$ as a first order system of the form

$$\dot{x}^j = \hat{x}^j, \quad \dot{\hat{x}}^j = -\Gamma^j_{km}(x)\,\hat{x}^k\hat{x}^m, \qquad (1 \le j \le n). \tag{12}$$

Now the right hand side of (12) is given by the L^∞_{loc}-vector field

$$f(x^1, \hat{x}^1, \ldots, x^n, \hat{x}^n) = (\hat{x}^1, -\Gamma^1_{km}(x)\hat{x}^k\hat{x}^m, \ldots, \hat{x}^n, -\Gamma^n_{km}(x)\hat{x}^k\hat{x}^m)$$

defined on some open subset U of $\mathbb{R}^{2n}$. On U we set $N := \{x^1 = 0\}$ and $D^\pm := \{\pm x^j > 0\}$. Then the unit normal of N pointing from D^- to D^+ is e_1, i.e., the first standard unit vector.

IOP Conf. Series: Journal of Physics: Conf. Series **968** (2018) 012010 doi:10.1088/1742-6596/968/1/012010

It follows that the projection of the limits of $f|_{D^\pm}$ on N onto the normal coincide and are just given by $f_N^\pm = \hat{x}^1 = \dot{x}^1$. So Corollary 3.5 applies if $\dot{x}^1(t_0) \neq 0$ ($\dot{x}^1(t_0) > 0$ in this case) where t_0 is the parameter value when γ (first) hits $\mathcal{N}$, i.e., $x^1(t_0) = 0$. But this just means that γ does hit $\mathcal{N}$ transversally and we have the following result:

Proposition 3.6. (Sufficient conditions for uniqueness) Let (M, g) be a $C^{0,1}$-semi-Riemannian manifold with g smooth off a C^∞-hypersurface[1] $\mathcal{N}$. Then the (Filippov) geodesics starting in $M \setminus \mathcal{N}$ that hit $\mathcal{N}$ transversally are globally unique and they cross from $\mathcal{D}^-$ into $\mathcal{D}^+$ or vice versa.

This proposition especially applies if the hypersurface $\mathcal{N}$ is *totally geodesic* in the following sense: Every (Filippov) geodesic starting in $\mathcal{N}$ tangentially to $\mathcal{N}$ remains initially in $\mathcal{N}$. Note that the notion of a totally geodesic submanifold in low regularity becomes somewhat subtle. In particular, in the situation at hand the second fundamental form will generically be not defined on all of $\mathcal{N}$.

Observe that the (Filippov) geodesics γ crossing $\mathcal{N}$ at $\gamma(t_0)$ coincide for all $t \neq t_0$ with the smooth geodesics of $\mathcal{D}^\pm$. Hence by their C^1-property $g(\dot\gamma, \dot\gamma)$ is globally constant and these geodesics do have a causal character. Moreover they are extremal curves by the following argument which we only detail in the Riemannian case: Suppose there is a curve λ connecting $\gamma(t_1) \in \mathcal{D}^-$ with $\gamma(t_2) \in \mathcal{D}^+$, which is shorter than $\gamma|_{[t_1,t_2]}$. Then set $t' = \sup\{t > t_1 : \gamma(t) = \lambda(t)\}$ and suppose $t' < t_0$. By continuity $\gamma(t') = \lambda(t')$ and we choose a totally normal neighbourhood $U \subseteq \mathcal{D}^-$ of $\gamma(t')$. Then λ has to be minimizing in U which contradicts the fact that λ is not the radial geodesic γ in U. So $\gamma(t) = \lambda(t)$ for all $t \neq t_0$ and hence everywhere by continuity.

Finally the above result suggests the idea to explicitly obtain the geodesics of M by appropriately matching the geodesics of each 'side' $\mathcal{D}^\pm$ across $\mathcal{N}$. We will discuss these matters in the closing section below.

3.4. The C^1-matching of geodesics

In this final section we discuss the matching of geodesics in locally Lipschitz semi-Riemannian manifolds with the metric smooth off a hypersurface $\mathcal{N}$. Indeed such an approach has been frequently applied in the literature on impulsive gravitational waves, see e.g. [6, 29, 30] and the references given in [27, Sec. 3.2]. The idea is to explicitly calculate the geodesics in $\mathcal{D}^\pm$ which is often possible (only) in coordinates which do not extend to the 'matching hypersurface' $\mathcal{N}$. Then one matches the geodesics γ^- of $\mathcal{D}^-$ that hit $\mathcal{N}$ 'from below' to the geodesics γ^+ of $\mathcal{D}^+$ that hit $\mathcal{N}$ 'from above' across $\mathcal{N}$. Explicitly in coordinates which cover $\mathcal{N}$ one sets $\gamma^-(t_0) = \gamma^+(t_0)$ and $\dot\gamma^-(t_0) = \dot\gamma^+(t_0)$ where t_0 is the parameter value where the respective geodesics hit $\mathcal{N}$. Obviously one needs to involve the derivatives to obtain the correct number of equations to match all the data and this is why one refers to this approach as C^1-matching procedure.

Often such calculations were done heuristically without supplying the necessary arguments which we collect here, cf. also [27, Rem. 4.1]. In fact the matching mathematically makes sense only if the following facts on the geodesics of M have been established:

(i) The geodesics reaching $\mathcal{N}$ cross it (rather than being reflected by or trapped into $\mathcal{N}$).

(ii) These geodesics are unique (rather than e.g. branching).

[1] In the sense specified above.

IOP Conf. Series: Journal of Physics: Conf. Series **968** (2018) 012010 doi:10.1088/1742-6596/968/1/012010

(iii) These geodesics are at least of C^1-regularity.

Indeed we can extract all this necessary information from the discussion in the previous section to obtain the following result:

Corollary 3.7. (The C^1-matching) Let (M, g) be a $C^{0,1}$-semi-Riemannian manifold with g smooth off a smooth hypersurface[1] $\mathcal{N}$. Let $\gamma^- : (a, b] \to \mathcal{D}^- \cup \mathcal{N}$ be a smooth geodesic with $\gamma(b) \in \mathcal{N}$ and $\dot\gamma(b) \notin T_{\gamma(b)}\mathcal{N}$. Then there exists $c > b$ and a unique (Filippov) geodesic $\gamma\colon (a, c) \to M$ such that $\gamma|_{(a,b]} = \gamma^-$.

Now the C^1-matching of the data can be made explicitly by defining $\gamma^+ := \gamma|_{[b,c)}$. Indeed we then have $\gamma^-(b) = \gamma^+(b)$ and $\dot\gamma^-(b) = \dot\gamma^+(b)$. Also we remark that this procedure allows to derive the (Filippov) geodesics crossing $\mathcal{N}$ simply by matching the smooth 'background' geodesics on either side in a C^1-manner without the need to go into the details of Filippov's theory.

This holds true even in case one needs to invoke the fact that $\mathcal{N}$ is totally geodesic to rule out that geodesics hit $\mathcal{N}$ tangentially. Indeed this fact can often be derived purely from knowledge of the 'background geodesics' $\gamma^\pm$ and the C^1-regularity of the (Filippov) geodesics cf. [27, Sec. 3.6].

Finally in case $\mathcal{N}$ fails to be totally geodesic (as is the case for all classes of expanding impulsive gravitational waves) one might still make use of Proposition 3.6 and Corollary 3.7 (and hence establish the $C^{1,1}$-matching) by showing 'by hand' that the 'background geodesics' do meet $\mathcal{N}$ transversally, cf. [28, Sec. 3.3].

4. Outlook and open problems

We conclude with listing some open question and discuss further lines of research.

- Riemannian du Bois-Reymond-trick for Lipschitz continuous metrics
  Let g be a Lipschitz continuous Riemannian metric and let γ be a Lipschitz continuous minimizer. Is there a way to see that γ has to be C^2 and that it satisfies the geodesic equations in the sense of Filippov?
- Lorentzian du Bois-Reymond trick
  Let g be a C^1 Lorentzian metric and let γ be a Lipschitz continuous maximizer between timelike related points. Does γ have to be timelike? This would imply that one could apply the du Bois-Reymond trick to get that γ is C^2 and that it satisfies the geodesic equations.
- Regularity of maximal causal curves
  Is there an analogue of the result by Lytchak and Yaman [21] for Lorentzian Hölder or Lipschitz continuous metrics? Even for C^1 metrics the regularity of maximal curves is unclear (cf. the point above).
- Causal character of maximizing causal curves
  What is the minimal regularity of a Lorentzian metric to ensure that maximal causal curves have a causal character? By Example 2.6 the metric has to be at least Lipschitz continuous[2].

[1] In the sense specified above.
[2] Note added in proof: In [10] it was meanwhile shown that $g \in C^{0,1}$ is actually sufficient.

- Properties of Filippov geodesics
  Can one say more about Filippov geodesics in Riemannian and Lorentzian signature? Do they have to be minimizing or maximizing, respectively, in some sense? Observe that it is not expected that Lorentzian Filippov geodesics have a causal character.

Acknowledgments

The authors are very grateful for the hospitality during the conference in Florence and want to especially thank Ettore Minguzzi for the friendly organization of that meeting. We also thank Michael Kunzinger, Jiří Podolský and Robert Švarc for constantly sharing their expertise. This work was supported by the Austrian Science Fund FWF, grants P28770 and P26859.

References

[1] O. Bolza. *Lectures on the calculus of variations.* Chicago: University of Chicago Press. XVI u. 271 S. 8°, 1904.
[2] D. Burago, Y. Burago, and S. Ivanov. *A course in metric geometry*, volume 33 of *Graduate Studies in Mathematics*. American Mathematical Society, Providence, RI, 2001.
[3] A. Y. Burtscher. Length structures on manifolds with continuous Riemannian metrics. *New York J. Math.*, 21:273–296, 2015.
[4] P. T. Chruściel and J. D. E. Grant. On Lorentzian causality with continuous metrics. *Classical Quantum Gravity*, 29(14):145001, 32, 2012.
[5] J. Cortés. Discontinuous dynamical systems: a tutorial on solutions, nonsmooth analysis, and stability. *IEEE Control Syst. Mag.*, 28(3):36–73, 2008.
[6] V. Ferrari, P. Pendenza, and G. Veneziano. Beam-like gravitational waves and their geodesics. *Gen. Relativity Gravitation*, 20(11):1185–1191, 1988.
[7] A. F. Filippov. *Differential equations with discontinuous righthand sides*, volume 18 of *Mathematics and its Applications (Soviet Series)*. Kluwer Academic Publishers Group, Dordrecht, 1988. Translated from the Russian.
[8] G. J. Galloway, E. Ling, and J. Sbierski. Timelike completeness as an obstruction to C^0-extensions. *Commun. Math. Phys*, online first, 2017. arXiv:1704.00353[gr-qc].
[9] M. Graf, J. D. E. Grant, M. Kunzinger, and R. Steinbauer. The Hawking-Penrose singularity theorem for $C^{1,1}$-Lorentzian metrics. *Commun. Math. Phys*, online first, 2017. arXiv:1706.08426[math-ph].
[10] M. Graf and E. Ling. Maximizers in Lipschitz spacetimes are either timelike or null. *preprint*, 2017. arXiv:1712.06504v1[gr-qc].
[11] S. Haller. *Generalized regularity and solution concepts for differential equations.* PhD thesis, University of Vienna, 2008. arXiv:0806.1451[math.AP].
[12] P. Hartman. On the local uniqueness of geodesics. *Amer. J. Math.*, 72:723–730, 1950.
[13] P. Hartman and A. Wintner. On the problems of geodesics in the small. *Amer. J. Math.*, 73:132–148, 1951.
[14] D. Hilbert. Über das Dirichletsche Prinzip. *Math. Ann.*, 59(1-2):161–186, 1904.
[15] M. W. Hirsch. *Differential topology*, volume 33 of *Graduate Texts in Mathematics*. Springer-Verlag, New York, 1994. Corrected reprint of the 1976 original.
[16] M. Kunzinger and C. Sämann. Lorentzian length spaces. *preprint*, 2017. arXiv:1711.08990[math.DG].
[17] M. Kunzinger, R. Steinbauer, and M. Stojković. The exponential map of a $C^{1,1}$-metric. *Differential Geom. Appl.*, 34:14–24, 2014.
[18] M. Kunzinger, R. Steinbauer, M. Stojković, and J. A. Vickers. A regularisation approach to causality theory for $C^{1,1}$-Lorentzian metrics. *Gen. Relativity Gravitation*, 46(8):Art. 1738, 18, 2014.
[19] M. Kunzinger, R. Steinbauer, M. Stojković, and J. A. Vickers. Hawking's singularity theorem for $C^{1,1}$-metrics. *Classical Quantum Gravity*, 32(7):075012, 19, 2015.
[20] M. Kunzinger, R. Steinbauer, and J. A. Vickers. The Penrose singularity theorem in regularity $C^{1,1}$. *Classical Quantum Gravity*, 32(15):155010, 12, 2015.
[21] A. Lytchak and A. Yaman. On Hölder continuous Riemannian and Finsler metrics. *Trans. Amer. Math. Soc.*, 358(7):2917–2926, 2006.
[22] E. Minguzzi. Limit curve theorems in Lorentzian geometry. *J. Math. Phys.*, 49(9):092501, 18, 2008.

[23] E. Minguzzi. Convex neighborhoods for Lipschitz connections and sprays. *Monatsh. Math.*, 177(4):569–625, 2015.

[24] E. Minguzzi and M. Sánchez. The causal hierarchy of spacetimes. In *Recent developments in pseudo-Riemannian geometry*, ESI Lect. Math. Phys., pages 299–358. Eur. Math. Soc., Zürich, 2008.

[25] B. O'Neill. *Semi-Riemannian geometry*, volume 103 of *Pure and Applied Mathematics*. Academic Press, Inc. [Harcourt Brace Jovanovich, Publishers], New York, 1983. With applications to relativity.

[26] B. E. Paden and S. S. Sastry. A calculus for computing Filippov's differential inclusion with application to the variable structure control of robot manipulators. *IEEE Trans. Circuits and Systems*, 34(1):73–82, 1987.

[27] J. Podolský, C. Sämann, R. Steinbauer, and R. Švarc. The global existence, uniqueness and C^1-regularity of geodesics in nonexpanding impulsive gravitational waves. *Classical Quantum Gravity*, 32(2):025003, 23, 2015.

[28] J. Podolský, C. Sämann, R. Steinbauer, and R. Švarc. The global uniqueness and C^1-regularity of geodesics in expanding impulsive gravitational waves. *Classical Quantum Gravity*, 33(19):195010, 23, 2016.

[29] J. Podolský and R. Steinbauer. Geodesics in spacetimes with expanding impulsive gravitational waves. *Phys. Rev. D (3)*, 67(6):064013, 13, 2003.

[30] J. Podolský and R. Švarc. Refraction of geodesics by impulsive spherical gravitational waves in constant-curvature spacetimes with a cosmological constant. *Phys. Rev. D(3)*, 81(12):124035, 19, 2010.

[31] F. Poupaud and M. Rascle. Measure solutions to the linear multi-dimensional transport equation with non-smooth coefficients. *Comm. Partial Differential Equations*, 22(1-2):337–358, 1997.

[32] C. Sämann. Global hyperbolicity for spacetimes with continuous metrics. *Ann. Henri Poincaré*, 17(6):1429–1455, 2016.

[33] J. Sbierski. The C^0-inextendibility of the Schwarzschild spacetime and the spacelike diameter in Lorentzian geometry. *Journal of Differential Geometry*, to appear, arXiv:1507.00601[gr-qc], 2017.

[34] R. Steinbauer. Every Lipschitz metric has C^1-geodesics. *Classical Quantum Gravity*, 31(5):057001, 3, 2014.

IOP Conf. Series: Journal of Physics: Conf. Series **968** (2018) 012011 doi:10.1088/1742-6596/968/1/012011

Green operators for low regularity spacetimes

Yafet Sanchez Sanchez and James Vickers

Mathematical Sciences and STAG Research Centre, University of Southampton, Southampton, SO17 1BJ.

E-mail: Y.SanchezSanchez@soton.ac.uk and J.A.Vickers@soton.ac.uk

Abstract. In this paper we define and construct advanced and retarded Green operators for the wave operator on spacetimes with low regularity. In order to do so we require that the spacetime satisfies the condition of generalised hyperbolicity which is equivalent to well-posedness of the classical inhomogeneous problem with zero initial data where weak solutions are properly supported. Moreover, we provide an explicit formula for the kernel of the Green operators in terms of an arbitrary eigenbasis of H^1 and a suitable Green matrix that solves a system of second order ODEs.

1. Introduction

The existence of Green operators for normally hyperbolic operators is well understood for globally hyperbolic spacetimes where the spacetime metric is smooth [1]. However, there are two important motivations for analysing spacetime metrics with finite differentiability. Firstly, there are several models of physical phenomena that require finite metric regularity. These include impulsive gravitational waves, stars with well-defined surfaces, general relativistic fluids and cosmic strings. Secondly, Einstein's equations, viewed as a system of hyperbolic PDEs, can be naturally formulated in function spaces with finite regularity.

When the spacetime is of finite regularity the existence of Green operators for the Klein-Gordon equation has been explored using semigroup techniques in the Hamiltonian formalism [2, 3, 4]. We complement such an analysis by providing a definition of Green operators subject only to the well posedness of the classical problem. Furthermore we show how to explicitly construct both the advanced and retarded the Green operators needed in order to formulate quantum field theory on a non-smooth background.

In the first section of the paper we give the definition of the advanced and retarded Green operators in the smooth case and define the concept of generalised hyperbolicity. In the second section of the paper we extend the definitions to the non-smooth setting and prove the existence of the Green operators for low regularity spacetimes. In the third section of the paper we give an explicit formula for the advanced Green operator in the non-smooth setting.

2. Preliminaries

2.1. The general setting

We will consider a smooth manifold M of the form $M = (0,T)\times\Sigma$ and its closure $\overline{M} = [0,T]\times\Sigma$ where Σ is a closed compact smooth manifold. Given a smooth Lorentzian metric g_{ab} of the form

$$ds^2 = N^2dt^2 - \gamma_{ij}(dx^i + \beta^i dt)(dx^j + \beta^j dt) \tag{2.1}$$

IOP Conf. Series: Journal of Physics: Conf. Series **968** (2018) 012011 doi:10.1088/1742-6596/968/1/012011

the wave operator $\Box_g$ acting on a scalar function u is given by

$$\Box_g u = \frac{1}{N\sqrt{\gamma}}\left(\partial_t\left(N\sqrt{\gamma}\frac{1}{N^2}\partial_t u\right)\right)$$
$$+\frac{1}{N\sqrt{\gamma}}\left(\partial_t\left(N\sqrt{\gamma}\frac{\beta^i}{N^2}\partial_i u\right)+\partial_j\left(N\sqrt{\gamma}\frac{\beta^j}{N^2}\partial_t u\right)\right) \tag{2.2}$$
$$-\frac{1}{N\sqrt{\gamma}}\partial_i\left(N\sqrt{\gamma}(\gamma^{ij}-\frac{\beta^i\beta^j}{N^2})\partial_j u\right)$$

where N is the lapse function, β^i the shift, γ_{ij} the metric on Σ and $\sqrt{\gamma}dx$ the induced volume form on Σ. We use a, b, c, d etc. to denote spacetime indices and i, j, k etc. to denote purely spatial indices.
The advanced zero initial data inhomogeneous problem for the wave equation on $\overline{M}$ is given by

$$\Box_g u = f \text{ on } M \tag{2.3}$$
$$u(0,x) = 0 \text{ on } \Sigma_0 = \Sigma \times \{0\} \tag{2.4}$$
$$\dot{u}(0,x) = 0 \text{ on } \Sigma_0 = \Sigma \times \{0\} \tag{2.5}$$

where $f : M \to \mathbb{R}$ is a smooth function. Similarly the retarded zero initial data inhomogeneous problem for the wave equation on $\overline{M}$ is given by

$$\Box_g u = f \text{ on } M \tag{2.6}$$
$$u(T,x) = 0 \text{ on } \Sigma_T = \Sigma \times \{T\} \tag{2.7}$$
$$\dot{u}(T,x) = 0 \text{ on } \Sigma_T = \Sigma \times \{T\} \tag{2.8}$$

where $f : M \to \mathbb{R}$ is a smooth function.
If the spacetime (M, g_{ab}) is globally hyperbolic the advanced and retarded zero initial data inhomogeneous problems are both well posed i.e. there exist unique solutions $u \in C^\infty(M)$ which depend continuously on f [1]. Moreover, in the globally hyperbolic case C^∞ well-posedness is equivalent to the existence of unique advanced and retarded Green operators [1]. For convenience we recall the definition of these.

Definition 2.1. A linear map $E^+ : D(M) \to C^\infty(M)$ satisfying

- $\Box_g E^+ = id_{D(M)}$
- $E^+\Box_g|_{D(M)} = id_{D(M)}$
- $supp(E^+\psi) \subset J^+(supp(\psi))$ for all $\psi \in D(M)$

is called an advanced Green operator for $\Box_g$. The retarded Green operators E^- are defined in a similar fashion.

For a smooth globally hyperbolic spacetime the splitting theorem says that the metric may be written in the form

$$ds^2 = +N^2dt^2 - \gamma_{ij}dx^i dx^j \tag{2.9}$$

and this remains true even in low regularity [5]. The aim of the paper is to define such advanced and retarded Green operators for the metrics of the above form where the lapse function N and the induced Riemannian metric γ_{ij} are no longer smooth.

IOP Conf. Series: Journal of Physics: Conf. Series **968** (2018) 012011 doi:10.1088/1742-6596/968/1/012011

Notation. We denote the derivative of a function u with respect t by u_t or $\partial_t u$ and u_i or $\partial_i u$ if it is with respect to the other x^i-coordinates. The space of smooth functions of compact support will be denoted by $D(M)$. A function f on an open set $\mathcal{U}$ of $\mathbb{R}^n$ is said to be Lipschitz if there is some constant K such that for each pair of points $p, q \in \mathcal{U}$, $|f(p) - f(q)| \leq K|p - q|$, where $|p|$ denotes the usual Euclidean distance. We denote by $C^{k,1}$ those C^{k-1} functions where the k-derivative is a Lipschitz function. A function f on M is said to be Lipschitz or $C^{k,1}$ if f is Lipschitz or $C^{k,1}$ in some coordinate chart.
In the analysis below we will be working with spaces such as $L^2(\Sigma)$, $H^1(\Sigma)$ $H^{-1}(\Sigma)$ which are defined with respect to a smooth background Riemannian metric h_{ij} on Σ with ν_h the corresponding volume form. We define $L^2(\Sigma)$ to be the space of real valued functions g on Σ such that $\int_\Sigma g^2 \nu_h < \infty$ and we denote the associated inner product by $(f, g)_{L^2} = \int_\Sigma fg\nu_h$. The space $H^k(\Sigma)$ are the real valued functions g such that their first k derivatives are in $L^2(\Sigma)$ and the space $H^{-k}(\Sigma)$ are the bounded linear functionals on $H^k(\Sigma)$. Note that a different choice of background metric simply results in an equivalent norm on all these spaces.
We also define the space $L^2(M, g)$ to be the space of real valued functions ϕ on M such that $\int_M \phi^2 \nu_g < \infty$ which is defined with respect the volume form given by the metric (2.9) and $L^2(\Sigma_t, \gamma)$ to be the space of real valued functions ψ on $\Sigma_t = \{t\} \times \Sigma$ for $0 \leq t < T$ such that $\int_{\Sigma_t} \psi^2 \nu_\gamma < \infty$ which is defined with respect the volume form given by the Riemannian metric γ_{ij} that appears in (2.9) . We denote in a similar way the Sobolev space $H^k(M, g)$ and $H^k(\Sigma, \gamma)$.
We will also often think of a function $\mathbf{v}(t, x)$ as a map from $[0, T]$ to a function $\mathbf{v}(t)(\cdot)$ in some Hilbert space $X(\Sigma)$ given by $\mathbf{v}(t)(x) = \mathbf{v}(t, x)$. For example $L^2(0, T; X(\Sigma))$ is the space of functions

$$\mathbf{v} : [0, T] \to X(\Sigma) \tag{2.10}$$
$$t \mapsto \mathbf{v}(t) \tag{2.11}$$

such that $\mathbf{v}(t) \in X(\Sigma)$ and

$$\left(\int_0^T ||\mathbf{v}(t)||^2_{X(\Sigma)} dt \right)^{\frac{1}{2}} < \infty \tag{2.12}$$

When thinking of $\mathbf{v}$ in this way, we will denote the time derivative by $\dot{\mathbf{v}}$.
We will also use the space $C^k([0, T]; X(\Sigma))$ which is the space of functions

$$\mathbf{v} : [0, T] \to X(\Sigma) \tag{2.13}$$
$$t \mapsto \mathbf{v}(t) \tag{2.14}$$

such that $\mathbf{v}(t) \in X(\Sigma)$ and

$$\max_{j=0,\ldots,k} \sup_{t \in [0,T]} ||\partial_{t^j} \mathbf{v}(t)||_{X(\Sigma)} < \infty \tag{2.15}$$

where $\partial_{t^l} \mathbf{v}(t)$ is the l-th derivative with respect time of $\mathbf{v}(t)$.

2.2. Generalised hyperbolicity

To prove the existence of Green operators in the non-smooth setting we will require that the spacetime satisfy the condition of *generalised hyperbolicity* (in H^1) which is essentially the requirement that solutions to the wave equation are well-posed [6], have the appropiate causal support and that the energy momentum of the solutions (see 2.23) is at least integrable.
In the low regularity setting the pointwise equation (2.3) may not make sense and therefore a notion of a weak solution is required.

IOP Conf. Series: Journal of Physics: Conf. Series **968** (2018) 012011 doi:10.1088/1742-6596/968/1/012011

Taking into account the form of the metric (2.9) and the geometric conditions that appear in Theorem 2.10 (see below) the wave operator may be written as

$$\Box_g u = \frac{\ddot{u}}{\gamma} - \frac{Lu}{\gamma} \tag{2.16}$$

where $-L$ is an elliptic operator in divergence form given by:

$$-Lu = -(\gamma^{ij}\gamma u_j)_i. \tag{2.17}$$

We can associate with the operator $-L$ the following bi-linear form given by:

$$B[u, v; t] := \int_\Sigma \gamma^{ij}(x, t)\gamma(x, t)u_i(x)v_j(x)dx^n \tag{2.18}$$

Using this bi-linear form, we make the following definition of an advanced weak solution.

Definition 2.2. We say a function:

$$u \in L^2([0, T]; H^1(\Sigma)), \text{ with } \dot{u} \in L^2([0, T]; L^2(\Sigma)), \ddot{u} \in L^2([0, T]; H^{-1}(\Sigma))$$

is an advanced weak solution of the zero initial data inhomogeneous problem (2.3), (2.4), (2.5) provided that:

(i) For each $v \in L^2([0, T]; H^1(\Sigma))$,

$$\int_0^T < \ddot{u}, v > dt + \int_0^T B[u, v; t]dt = (f, v)_{L^2(M,g)} \tag{2.19}$$

where $< \cdot, \cdot >$ denotes the dual pairing between the $H^{-1}(\Sigma)$ and $H^1(\Sigma)$ Sobolev spaces.

(ii) $u(0, x) = 0, \quad \dot{u}(0, x) = 0$

The definition of the retarded weak solution is analogous.

Remark 2.3. Notice that in our convention we have chosen to call advanced solutions those solutions that evolve from an initial time t' to a future time $t \geq t'$ and retarded if the solutions evolve in the opposite direction. We have used this convention as advanced solutions (in our sense) will be used to construct the advanced Green functions. However, there are other conventions for advanced and retarded functions in the literature with the ambiguity stemming from the fact that many authors use retarded fundamental solutions of the adjoint operator to construct the advanced Green operators. (see e.g [1] equation (3.8)).

Remark 2.4. The regularity of the weak solution implies that $u \in C([0, T], L^2(\Sigma))$ and $\dot{u} \in C([0, T], H^{-1}(\Sigma))$ (see [7]). Consequently, equation (ii) has to be understood in this sense.

Remark 2.5. The regularity needed to make sense of Equation (2.19) is that

$$u \in L^2([0, T]; H^1(\Sigma)), \text{ with }, \ddot{u} \in L^2([0, T]; H^{-1}(\Sigma)).$$

Therefore for every $u \in L^2([0, T]; H^1(\Sigma))$ with $\ddot{u} \in L^2([0, T]; H^{-1}(\Sigma))$ we can define the bounded linear functional $(\Box_g u)[\cdot]$ on $L^2([0, T]; H^1(\Sigma))$ given by

$$(\Box_g u)[v] = \int_0^T < \ddot{u}, v > dt + \int_0^T B[u, v; t]dt. \tag{2.20}$$

for all $v \in L^2([0, T]; H^1(\Sigma))$ and therefore the pairing $\langle \Box_g u, v \rangle_{L^2H^{-1}, L^2H^1}$ makes sense.

We say an advanced weak solution is regular if it satisfies a suitable energy estimate.

IOP Conf. Series: Journal of Physics: Conf. Series **968** (2018) 012011 doi:10.1088/1742-6596/968/1/012011

Definition 2.6. **(Regular Weak Solution)**
An advanced weak solution is **regular** if u satisfies the energy estimate

$$||u(t,\cdot)||_{L^2([0,T];H^1(\Sigma))} + ||\dot{u}(t,\cdot)||_{L^2([0,T];L^2(\Sigma))} + ||\ddot{u}(t,\cdot)||_{L^2([0,T];H^{-1}(\Sigma))} \leq C||f||_{L^2(M,g)} \tag{2.21}$$

We will also require the weak solutions to satisfy certain support properties. In the smooth case the condition of global hyperbolicity is such that a solution u of the inhomogeneous problem satisfies $supp(u) \subset J^+(supp(f)) \cup J^-(supp(f))$. We use this observation and the definition of the support of a distribution u i.e. $supp(u) = \{x \in M|$ for all neighbourhoods U of x there exist $\varphi \in D(M)$ with $supp(\varphi) \subset U$ and $u[\varphi] \neq 0\}$ to make the following definition:

Definition 2.7. **(Causally supported Weak Solution)** We say a weak solution is **causally supported to the future** if u satisfies the support condition

$$supp(u) \subset J^+(supp(f)) \tag{2.22}$$

in the sense of distributions. The definition of causally supported to the past is analogous. We say a weak solution is **causally supported** if it is causally supported to both the past and future i.e. $supp(u) \subset J^+(supp(f)) \cup J^-(supp(f))$.

We are now in a position to extend the notion of generalised hyperbolicity to the case where the wave equation only admits weak solutions.

Definition 2.8. **(Generalised hyperbolicity)** We say a spacetime M satisfies the advanced generalised hyperbolicity condition if the wave equation is well-posed in the following sense: There exists a unique regular weak solution u of $\Box_g u = f$ in the sense of definition 2.2 and definition 2.6 with homogeneous initial conditions

(i) $u(0,\cdot) = u|_{\Sigma_0} = 0$
(ii) $\dot{u}(0,\cdot) = \frac{\partial u}{\partial t}|_{\Sigma_0} = 0$

that is causally supported to the future. With the obvious modifications we define the condition of retarded generalised hyperbolicity. If a spacetime satisfies the advanced and retarded condition of generalised hyperbolicity we say the spacetime satisfies the condition of generalised hyperbolicity.

Remark 2.9. In the case of a smooth metric, global hyperbolicity is sufficient to guarantee that the condition of generalised hyperbolicity in the above sense is also satisfied.
On the other hand even in the smooth case the notion of generalised hyperbolicity is more general. For example a spacetime with timelike boundary cannot be globally hyperbolic, but the initial value problem with reflecting timelike boundary conditions can be well-posed [8].

The following series of theorems give sufficient conditions to ensure such that a low regularity spacetime (M, g) satisfies the condition of generalised hyperbolicity. For simplicity, we only deal with the condition of advanced generalised hyperbolicity. We first show the existence and uniqueness of regular weak solutions of the initial value problem under certain regularity assumptions.

Theorem 2.10. *Let the metric g_{ab} given by Equation 2.9 on M satisfy the following conditions*

1. $\gamma^{ij} \in C^1([0,T], L^\infty(\Sigma))$.
2. *The scalar coefficient of the volume form given by $\sqrt{\gamma}$ for the induced metric γ_{ij} is bounded from below by a positive real number, i.e., $|\sqrt{\gamma}| > \eta$ for some $\eta \in \mathbb{R}^+$*
3. *The lapse function N can be chosen as $N = \sqrt{\gamma}$.*

IOP Conf. Series: Journal of Physics: Conf. Series **968** (2018) 012011 doi:10.1088/1742-6596/968/1/012011

4. There exists a constant $\theta > 0$ such that

$$\sum_{i,j=1}^{n} \gamma^{ij}\gamma\xi_i\xi_j \geq \theta|\xi|^2$$

for all $(t,x) \in M, \xi \in \mathbb{R}^n$.

Then for every $f \in L^2(M,g)$ there exists a unique weak regular solution to the initial value problem (both forwards and backwards in time).

Remark 2.11. The regularity conditions on both the lapse and the three metric required in Theorem 2.10 depend on the splitting. However the conditions required for both of these are invariant under any diffeomorphism that preserves the splitting.

Proof of Theorem 2.10 The main idea is to use Galerkin's method. We sketch the proof below. For a complete proof see [7, 9, 10].

- Introduce an orthogonal basis in $H^1(\Sigma)$ $\{w_k\}_{k=1}^{\infty}$ which is also an orthonormal basis in $L^2(\Sigma)$. An example of such a basis is given by the eigenfunctions of the Laplace-Beltrami operator with respect to the background Riemannian metric h_{ij}.
- Insert the approximate solution $u^{(m)}(t,x) = \sum_{k=1}^{m} d_k^{(m)}(t)w_k(x)$ into the projected wave equation $(u_{tt}^{(m)}, w_k)_{L^2(\Sigma)} + B[u^{(m)}, w_k] = (\gamma f, w_k)_{L^2(\Sigma)}$ for all $k = 1, ..., m$. This gives a first order system of ODEs for $d_k^{(m)}(t)$ with zero initial data on $t = 0$.
- Prove that the $u^{(m)}$ satisfy the energy estimate uniformly in m. This is the crucial step for the Galerkin's method to work as it will provide sufficient control over the approximated solutions to converge to a weak solution. The uniform control is obtained by finding a bound in terms of the norms of a truncated expansion of the source term and initial data with respect the chosen basis. While these bounds are not independent of m both terms are uniformly bounded in term of the initial data and the source. See [7] for details.
- Apply Banach-Alouglu theorem to extract a subsequence which is a solution to the weak formulation
- Use lower semi-continuity of the norm and energy estimates to prove uniqueness, stability and regularity of the weak solution.

At this point we have shown that regular weak solutions to the initial value problem exist. However, we have not shown that the support of such solutions satisfies the causal support condition to the future. We therefore prove the theorem below, which guarantees the correct support of these solutions and establishes the required generalised hyperbolicity.
Notice that Theorem 2.12 requires additional regularity conditions compared to Theorem 2.10. It does however have the advantage that the regularity condition is invariant under any diffeomorphism, not just those that respect the splitting. It may be possible to relax the regularity conditions to those of Theorem 2.10 by analysing the causal structure of spacetimes with low regularity as in [5, 12, 13] and using results on propagation of singularities for differential equations with rough coefficients as in [16].

Theorem 2.12. *Let the metric g_{ab} be a $C^{1,1}$ globally hyperbolic spacetime metric on M and the source function f be an element of $H^1(M,g)$ then (M,g) satisfies the condition of generalised hyperbolicity.*

IOP Conf. Series: Journal of Physics: Conf. Series **968** (2018) 012011 doi:10.1088/1742-6596/968/1/012011

Proof of Theorem 2.12. As described above the results of [7, 9, 10] establish well-posedness. The only thing that remains to be shown is that the weak solutions have the correct support. To do so we follow [11] and obtain an energy inequality that determines the support of the solution. Consider a weak solution u with energy-momentum tensor

$$T^{ab}[u] = \left(g^{ac}g^{bd} - \frac{1}{2}g^{ab}g^{cd}\right)\frac{\partial u}{\partial x^c}\frac{\partial u}{\partial x^d} \tag{2.23}$$

Let K be a compact subset of Σ_t. Then if we contract the energy-momentum tensor with a Lipschitz timelike vector field, Υ_a and use the fact that $J^-_{\Sigma_t}(K) \cap J^+(\Sigma_0)$ is compact by causal theory for globally hyperbolic $C^{1,1}$ metrics [12, 13, 5] (where $J^-_{\Sigma_t}(K)$ denotes the causal past of a compact region $K \subset \Sigma_t$), then provided we have sufficient regularity that one can apply the divergence theorem we obtain

$$\int_{J^-_{\Sigma_t}(K)\cap J^+(\Sigma_0)} \operatorname{div}\left(T^{ab}\Upsilon_a\right)\nu_g = \int_{\partial(J^-_{\Sigma_t}(K)\cap J^+(\Sigma_0))} T^{ab}\Upsilon_a n_b \nu_\gamma \tag{2.24}$$

In order for the above equation to be well-defined we require that $\operatorname{div}(T^{ab}\Upsilon_a)$ should be integrable with respect the volume form ν_g, and this in turn requires that the weak solutions are sufficiently regular. In fact it is enough that the weak solutions have two derivatives in $L^2(M,g)$ if the metric and the timelike vector field are in the space $C^{0,1}$ (see [14]). This is ensured by the additional regularity of the metric $g_{ab} \in C^{1,1}$ and source $f \in H^1(M,g)$ required in the conditions of Theorem 2.12 compared to Theorem 2.10. Moreover, such an improvement in regularity also guarantees that u belongs to the space $C^0([0,T],H^1(\Sigma)) \cap C^1([0,T],L^2(\Sigma))$. See [7] for a proof of these results.
The left hand side of 2.24 takes the explicit form:

$$\int_{J^-_{\Sigma_t}(K)\cap J^+(\Sigma_0)} \left(\left(g^{ab}\frac{\partial u}{\partial x^b}\Upsilon_a\right)\Box_g u + T^{ab}\nabla_b\Upsilon_a\right)\nu_g \tag{2.25}$$

while the right hand side takes the form:

$$\left(\int_K + \int_{\Sigma_0}\right) T^{ab}\Upsilon_a n_b \nu_\gamma + \int_{\mathcal{H}} T^{ab}\Upsilon_a n_b \nu_h. \tag{2.26}$$

where $\mathcal{H} = \partial(J^-_{\Sigma_t}(K) \cap J^+(\Sigma_0))\backslash(\Sigma_0 \cup K)$, ν_h is the induced volume form on $\mathcal{H}$ and n_b denotes outward pointing normal vectors on K, Σ_0 and $\mathcal{H}$.
Moreover, in the globally hyperbolic $C^{1,1}$ case we have that $\mathcal{H}$ is a null hypersurface ruled by null geodesics [13] and by the dominant energy condition [11] which follows from the specific form of 2.23 we have that

$$\int_{\mathcal{H}} T^{ab}\Upsilon_a n_b \nu_h \geq 0 \tag{2.27}$$

For the integral over K we define the energy integral:

$$E_K(t) = \int_K T^{ab}\Upsilon_a n_b \nu_\gamma \tag{2.28}$$

and we may define a similar energy integral E_{Σ_0} for the initial surface Σ_0.
We now use the energy integral $E_K(t)$ to construct a H^1 type Sobolev norm on K defined by

$$\tilde{\|}u\tilde{\|}^1_K := \left[\int_K \left(\left(\frac{\partial u}{\partial t}\right)^2 + \sum_{i=1}^{3}\left(\frac{\partial u}{\partial x^i}\right)^2\right)\nu_\gamma\right]^{\frac{1}{2}} \tag{2.29}$$

IOP Conf. Series: Journal of Physics: Conf. Series **968** (2018) 012011 doi:10.1088/1742-6596/968/1/012011

These norms are equivalent since as shown by Wilson [15]

$$C_1 E_K(t) \leq (\tilde{\|}u\tilde{\|}^1_K)^2 \leq C_2 E_K(t) \tag{2.30}$$

for constants $C_1, C_2 \geq 0$. In a similar way we may relate the initial energy integral E_{Σ_0} to an equivalent Sobolev type norm $\tilde{\|}u\tilde{\|}^1_{\Sigma_0}$ on Σ_0.
We now use the fact that in terms of this norm we have

$$\|\phi\|_{H^1(J^-_{\Sigma_{t'}}(K)\cap J^+(\Sigma_0),g)} \leq C\left(\int_0^t \left(\tilde{\|}\phi\tilde{\|}^1_K\right)^2 dt'\right)^{\frac{1}{2}} \tag{2.31}$$

where C is a constant that depends on g_{ab} and the interval $[0,t]$ [15]. Estimating all the terms in (2.24) by using the Cauchy Schwarz inequality together with the regularity of the metric and solutions gives the inequality:

$$E_K(t) \leq E_{\Sigma_0}(0) + k_0(\|\Box_g u\|_{L^2(J^-_{\Sigma_t}(K)\cap J^+(\Sigma_0),g)})^2 + k_1(\|u\|_{H^1(J^-_{\Sigma_t}(K)\cap J^+(\Sigma_0),g)})^2 \tag{2.32}$$

where k_0, k_1 are positive constants that depend on the metric g_{ab}, the vector field Υ_a and the covariant derivative $\nabla_b\Upsilon_a$.
Now rewriting (2.32) using (2.30) and (2.31) we find

$$E_K(t) \leq E_{\Sigma_0}(0) + k_0(\|\Box_g u\|_{L^2(J^-_{\Sigma_t}(K)\cap J^+(\Sigma_0),\nu_g)})^2 + k_2\int_0^t E(t')dt' \tag{2.33}$$

Using Gronwall's inequality we obtain

$$E_K(t) \leq K_4\left(E_{\Sigma_0}(0) + (\|\Box_g u\|_{L^2(J^-_{\Sigma_t}(K)\cap J^+(\Sigma_0),g)})^2\right) \quad \text{for all } t \leq T \tag{2.34}$$

where K_4 is a positive constant that depends on the chosen finite time t, the metric g_{ab}, the vector field Υ_a and the covariant derivative $\nabla_b\Upsilon_a$.
Rewriting everything in terms of the Sobolev type norms we obtain the expression:

$$(\tilde{\|}u\tilde{\|}^1_K)^2 \leq A\left((\tilde{\|}u\tilde{\|}^1_{\Sigma_0})^2 + (\|\Box_g u\|_{L^2(J^-_{\Sigma_t}(K)\cap J^+(\Sigma_0),g)})^2\right) \tag{2.35}$$

for some constant A.
By definition we have that the advanced weak solution has vanishing initial data, therefore $\tilde{\|}u\tilde{\|}^1_{\Sigma_0} = 0$ and we obtain the energy estimate

$$(\tilde{\|}u\tilde{\|}^1_K)^2 \leq A(\|\Box_g u\|_{L^2(J^-_{\Sigma_t}(K)\cap J^+(\Sigma_0),g)})^2 \tag{2.36}$$

This establishes the fact that the norm (2.29) of an advanced solution u of the zero initial data inhomogeneous problem $\Box_g u = f$ on $K \subset \Sigma_t$ is bounded by the value of the source f in the region $J^-_{\Sigma_t}(K) \cap J^+(\Sigma_0)$.
Suppose now that p does not belongs to $J^+(supp(f))$. Then $J^-(p)$ does not intersect $supp(f)$ and therefore we can construct a compact set $K \subset \Sigma_{t'}$ that contains p for some t' such that $J^-_{\Sigma'_t}(K) \cap J^+(\Sigma_0)$ does not intersect $supp(f)$. This implies that the right hand side of (2.36) vanishes and $u = 0$ on K. Then, taking a possible smaller region $K' \subset K$ and a small time interval ϵ we can form a neighbourhood $U = \epsilon \times K'$ that contains p and where u vanishes. In this region we have that $u = 0$ and this implies that p is not in the support of u which gives the desired result.

IOP Conf. Series: Journal of Physics: Conf. Series **968** (2018) 012011 doi:10.1088/1742-6596/968/1/012011

Remark 2.13. The norm defined by (2.29) is not a true Sobolev type norm as it is misssing the zero order terms. However, in the compact setting one has available the Poincaré inequality [7] which guarantees the vanishing of u. Alternatively one can work with a modified energy-momentum tensor S^{ab} which contains an additional $u^2 g_{ab}$ term and is equivalent to a genuine Sobolev norm (see e.g. [15]).

3. Green operators in the non-smooth setting

In this section we define the notion of advanced and retarded Green operators in a weak sense. Notice that Definition 2.1 can not be used in a setting with finite differentiability because weak solutions are in general not smooth so that one cannot expect $E^+(f) \in C^\infty(M)$ for $f \in D(M)$. In fact, even if the metric is smooth but f is not in $D(M)$ one can not expect smooth global solutions. We therefore propose the following definitions suitable for settings with finite differentiability.

Definition 3.1. A bounded linear map $E^+ : L^2(M,g) \to L^2([0,T]; H^1(\Sigma))$ satisfying

- $\langle \Box_g E^+(f), v\rangle_{L^2H^{-1},L^2H^1} = (f,v)_{L^2(M,g)}$ for all $f \in D(M), v \in L^2([0,T]; H^1(\Sigma))$
- $\langle E^+\Box_g(f), v\rangle_{L^2H^{-1},L^2H^1} = (f,v)_{L^2(M,g)}$ for all $f \in D(M), v \in L^2([0,T]; H^1(\Sigma))$
- $supp(E^+(f)) \subset J^+(supp(f))$ for all $f \in D(M)$

is an advanced weak Green operator for $\Box_g$. A retarded weak Green operator E^- is defined similarly.

Remark 3.2. If g_{ab} is in $C^{0,1}$ and for all $f \in D(M)$ we have $E^\pm(f)$ in $H^2(M,g)$ then the first property can be restated as

$$(\Box_g E^\pm f, v)_{L^2(M,g)} = (f,v)_{L^2(M,g)} \tag{3.1}$$

for all $f \in D(M), v \in L^2([0,T]; H^1(\Sigma))$.

Remark 3.3. If for all $f \in D(M)$ we have $E^\pm(\Box_g f)$ in $L^2(M,g)$ then the second property can be restated as

$$(E^\pm \Box_g f, v)_{L^2(M,g)} = (f,v)_{L^2(M,g)} \tag{3.2}$$

for all $f \in D(M), v \in L^2([0,T]; H^1(\Sigma))$

We now show that the notion of generalised hyperbolicity is sufficient to establish the existence of weak Green operators.

Theorem 3.4. *Let (M,g) be a globally hyperbolic spacetime with $g_{ab} \in C^{1,1}$. Then on M there are unique advanced and retarded weak Green operators for $\Box_g$.*

Proof of Theorem 3.4. We will only show the existence of the advanced weak Green operator. The existence of the retarded weak Green operator follows from time reversal.
Define the linear map $E^+ : L^2(M,g) \to L^2([0,T]; H^1(\Sigma))$ which sends a source function f to the advanced regular weak solution u^+ of the advanced zero initial data inhomogeneous problem with source f . That such u^+ exist and is unique is a consequence of generalised hyperbolicity which follows from Theorem 2.12.
We first show that this map is well defined and continuous.
Let $f = g \in L^2(M,g)$, then by definition we have that $E^+(f) = u^+$ and $E^+(g) = v^+$ are regular weak solutions of the zero initial data inhomogeneous problem with source function $f = g$ causally supported to the future. By the condition of generalised hyperbolicity which guarantees uniqueness of solutions, we have $u^+ = v^+$ and therefore $E^+(f) = E^+(g)$. Therefore the map is well defined.

IOP Conf. Series: Journal of Physics: Conf. Series **968** (2018) 012011 doi:10.1088/1742-6596/968/1/012011

That the map is continuous follows directly from the regularity of solutions. We have

$$||E^+(f)||_{L^2([0,T];H^1(\Sigma))} = ||u||_{L^2([0,T];H^1(\Sigma))} \leq ||f||_{L^2(M,g)} \tag{3.3}$$

We next show that E^+ is a suitable right inverse in the sense of Definition 3.1. First notice that the regularity of the weak solutions given by the condition of generalised hyperbolicity allows us to define for every advanced weak solution u^+ the linear map $\Box_g u^+[\cdot]$. See remark 2.5. Therefore given $E^+(f) = u^+$, we have

$$\Box_g E^+(f)[v] =: \Box_g u^+[v] = \int_0^T < \ddot{u}^+, v > dt + \int_0^T B[u^+, v; t]dt = (f, v)_{L^2(M,g)} \tag{3.4}$$

which gives

$$\langle \Box_g E^+ f, v\rangle_{L^2H^{-1},L^2H^1} = (f, v)_{L^2(M,g)} \tag{3.5}$$

for all $f \in D(M), v \in L^2([0,T]; H^1(\Sigma))$
Now we prove a similar equality for $E^+\Box_g f[v]$ to show that it is a left inverse. First we notice that by definition we have that $u^+ = E^+(\Box_g f)$ is an advanced weak solution of the inhomogeneous problem with source $\Box_g f$ and therefore we have that

$$\int_0^T < \ddot{u}^+, v > dt + \int_0^T B[u^+, v; t]dt = (\Box f, v)_{L^2(M,g)} \tag{3.6}$$

At the same time using integration by parts we have that

$$\int_0^T < \ddot{f}, v > dt + \int_0^T B[f, v; t]dt = (\Box f, v)_{L^2(M,g)} \tag{3.7}$$

so f is also a weak solution. Moreover, using the smoothness of f and the assumed regularity of the metric we can obtain an energy estimate which proves that f is a regular weak solution (see [7]). Then u^+ and f are weak regular solutions to the zero initial data inhomogeneous problem with the same source. So by uniqueness of the solutions we have that $u^+ = f$ which guarantees that

$$\langle E^+\Box_g f, v\rangle_{L^2H^{-1},L^2H^1} = (f, v)_{L^2(M,g)} \tag{3.8}$$

for all $f \in D(M), v \in L^2([0,T]; H^1(\Sigma))$
Finally to analyse the support of $E^+(f)$ we use that the weak solutions are causally supported. We have that $supp(E^+(f)) = supp(u^+) \subset J^+(supp(f))$ for all $f \in D(M)$ which gives the desired result.

Remark 3.5. Notice that to prove the first two conditions of definition 3.1 we use Theorem 2.10 while the condition that the solution has the correct causal support is a consequence of Theorem 2.12.

Remark 3.6. It is pointed out in remark 3.4.9 in [1] that Green operators on compact spacetimes with smooth metrics cannot exist as that would imply that $\Box_g : D(M) \to D(M)$ is injective. The proof uses the fact that if $\Box_g \phi = 0$ and Green operators exist, then one has $\phi = E^\pm \Box_g \phi = E^\pm 0 = 0$ which shows injectivity. However, any constant function is in the kernel of $\Box_g$. In our case, if one tries to repeat such a result one has the following equalities $(\phi, v) = (E^\pm \Box_g \phi, v)_{L^2(M,g)} = (E^\pm 0, v)_{L^2(M,g)} = (0, v)_{L^2(M,g)} = 0$. By the density of $L^2([0,T]; H^1(\Sigma))$ in $L^2(M, g)$ we can conclude that $\phi = 0$. which recovers the non-existence result in our setting also and explains the choice $M = (0,T) \times \Sigma$.

IOP Conf. Series: Journal of Physics: Conf. Series **968** (2018) 012011 doi:10.1088/1742-6596/968/1/012011

We next show that the formal adjoint of E^+ is E^-.

Theorem 3.7. *Given $\chi \in D(M)$ and $\varphi \in D(M)$ we have that*

$$\int_M E^+(\chi)\varphi\nu_g = \int_M \chi E^-(\varphi)\nu_g \tag{3.9}$$

In order to show this result we need the following Lemma

Lemma 3.8. *Given $\chi, \varphi \in H^2(M, g)$ with $\chi(0, x) = \chi_t(0, x) = 0$ and $\varphi(T, x) = \varphi_t(T, x) = 0$ we have that*

$$\int_M \Box_g \chi\varphi\nu_g = \int_M \chi\Box_g\varphi\nu_g \tag{3.10}$$

Proof of Lemma 3.8. The proof follows straightforward from using integration by parts twice and using the boundary conditions given by the hypothesis.

Proof of Theorem 3.7.

First, notice that because the metric has regularity $C^{1,1}$ and $\chi, \varphi \in D(M)$ we have that $E^+(\chi), E^-(\varphi) \in H^2(M, g)$. [7].

Hence,

$$\begin{aligned}
\int_M E^+(\chi)\varphi\nu_g &= (E^+(\chi), \varphi)_{L^2(M,g)} &(3.11)\\
&= (E^+(\chi), \Box_g E^-(\varphi))_{L^2(M,g)} &(3.12)\\
&= (\Box_g E^+(\chi), E^-(\varphi))_{L^2(M,g)} &(3.13)\\
&= (\chi, E^-(\varphi))_{L^2(M,g)} &(3.14)\\
&= \int_M \chi E^-(\varphi)\nu_g &(3.15)
\end{aligned}$$

where the second and fourth equality follows from remark 3.2 while the third inequality used the initial data given by the advanced and retarded zero initial data inhomogeneous problem and Lemma 3.8.

In the smooth case it is often convenient to consider the Green operators as bi-distributions on the product space. We show that this result extends to the non-smooth setting.

Theorem 3.9. *The map $\mathbf{E}^+ : D(M) \otimes D(M) \to \mathbb{R}$ given by*

$$\mathbf{E}^+(fg) = (E^+(f), g)_{L^2(M,g)} \tag{3.16}$$

can be extended to a distribution on $M \times M$.

Proof of Theorem 3.9.

Given $f \otimes g \in D(M) \otimes D(M)$ we have

$$\begin{aligned}
\mathbf{E}^+(fg) &= (E^+(f), g)_{L^2(M,g)} &(3.17)\\
&\leq ||E^+(f)||_{L^2(M,g)}||g||_{L^2(M,g)} &(3.18)\\
&\leq ||E^+(f)||_{L^2([0,T],H^1(\Sigma))}||g||_{L^2(M,g)} &(3.19)\\
&\leq C_1||f||_{L^2(M,g)}||g||_{L^2(M,g)} &(3.20)\\
&\leq C_1||f \otimes g||_{L^2(M,g)} &(3.21)\\
&\leq C_2||f \otimes g||_{C^0(M)} &(3.22)
\end{aligned}$$

for some constants C_1, C_2 and where we have used Holder's inequality and that $E^+(f)$ is a regular solution. Now taking into account that $D(M) \otimes D(M)$ is dense in $D(M \times M)$ we can extend $\mathbf{E}^+(fg)$ to a unique continuous linear mapping in $D(M \times M)$. This proves the claim.

Remark 3.10. See [1] for a proof of this theorem in the smooth setting.

IOP Conf. Series: Journal of Physics: Conf. Series **968** (2018) 012011 doi:10.1088/1742-6596/968/1/012011

4. Galerkin type approximation of E^+

In this section we will provide a construction of the operator E^+ and give a formula for the kernel. To construct the map E^+ we will build a sequence of operators E^+_m that converge in the weak operator topology to E^+. We will construct the operators E^+_m using Galerkin's method following the methods of [7, 9, 10].

First consider a complete orthogonal basis $\{w_j(x)\}_{j\in J}$ in $H^1(\Sigma)$ which is also orthonormal in $L^2(\Sigma)$. Then inserting the m-approximate solution $u^{(m)^+}(t,x) = \sum_{k=1}^m d_k^{(m)}(t)w_k(x)$ into the equation 2.19 we find that the functions $d_k^{(m)}$ for $k = 1, ..., m$ satisfies the system of second order ODEs

$$\ddot{d}^k_m(t) + \sum_{l=1}^m e^{kl}(t)d^l_m(t) = f^k(t) \tag{4.1}$$

where $e^{kl}(t) = B[w_k, w_l; t]$ are C^1 and symmetric and $f^k(t) = \int_\Sigma f(t,y)w_j(y)\gamma(y,s)dy$. (See Theorem 2.10)

We may find a Green matrix $F^{kj}_m(t,s)$ [17] for the initial value problem for this system of second order ODEs and write the solution in the form

$$d^k_m(t) = \int_0^T \sum_j F^{kj}_m(t,s)f^j(s)ds \tag{4.2}$$

So that

$$u^m(t,x) = \int_0^T \sum_{j,k} F^{kj}_m(t,s)f^j(s)w_k(x)ds \tag{4.3}$$

Recalling that $f^j(s) = \int_\Sigma f(s,y)w_j(y)\gamma(y,s)dy$ we have

$$u^m(t,x) = \int_0^T \int_\Sigma \sum_{j,k} F^{kj}_m(t,s)w_k(x)w_j(y)f(s,y)\gamma(y,s)dyds \tag{4.4}$$

We may now define the m-approximate advanced Green operator, E^+_m on $L^2(M,g)$ by

$$E^+_m(f) := \int_0^T \int_\Sigma \sum_{j,k} F^{kj}_m(t,s)w_k(x)w_j(y)f(s,y)\gamma(y,s)dyds \tag{4.5}$$

where $F^{kj}_m(t,s)$ is the Green matrix for the (regular) ODE problem.

The next step is to show that $\{u^m(t,x)\}_m$ is bounded uniformly in $L^2([0,T];H^1(\Sigma))$. This is a consequence of the energy estimates. See [7] for a derivation of the estimates and the uniform bound. Applying the Banach-Alaoglu theorem to the bounded sequence we can obtain a subsequence $\{u^{m_l}(t,x)\}_l$ that converges weakly to u^+ in $L^2([0,T];H^1(\Sigma))$ i.e. there is a subsequence $\{u^{m_l}\}_l$ that for every linear bounded operator ϕ on $L^2([0,T];H^{-1}(\Sigma))$ we have $\lim_{l\to\infty}\langle\phi, u^{m_l}\rangle_{L^2H^{-1},L^2H^1} = \langle\phi, u^+\rangle_{L^2H^{-1},L^2H^1}$.

Notice that the weak convergence of the sequence of functions $\{u^{m_l}\}$ is equivalent to convergence of the operators $\{E^+_{m_l}(f)\}$ in the weak operator topology. By definition 4.5 we have

$$\langle\phi, u^{m_l}\rangle_{L^2H^{-1},L^2H^1} = \langle\phi, E^+_{m_l}(f)\rangle_{L^2H^{-1},L^2H^1}$$

for all $\phi \in (L^2([0,T];H^{-1}(\Sigma)))$. Now by weak convergence of u^{m_l} to u^+ we have that $\lim_{l\to\infty}\langle\phi, u^{m_l}\rangle_{L^2H^{-1},L^2H^1} = \langle\phi, u^+\rangle_{L^2H^{-1},L^2H^1}$ for all $\phi \in L^2([0,T];H^{-1}(\Sigma))$ but this is equivalent to

$$\lim_{l\to\infty}\langle\phi, E^+_{m_l}(f)\rangle_{L^2H^{-1},L^2H^1} = \langle\phi, E^+(f)\rangle_{L^2H^{-1},L^2H^1} = \langle\phi, u^+\rangle_{L^2H^{-1},L^2H^1} \tag{4.6}$$

for all $\phi \in L^2([0,T];H^{-1}(\Sigma))$ which establishes the weak convergence in the operator topology. We may therefore write for every $v \in (L^2([0,T];H^1(\Sigma)) \subset L^2([0,T];H^{-1}(\Sigma))$

$$\langle v, u^+\rangle_{L^2H^{-1},L^2H^1} = \langle v, E^+(f)\rangle_{L^2H^{-1},L^2H^1} \tag{4.7}$$
$$= \lim_{l\to\infty}\int_M E^+_{m_l}(f)(x,t)v(t,x)\gamma(x,t)dtdx \tag{4.8}$$
$$= \lim_{l\to\infty}\int_M\int_M \sum_{j,k} F^{kj}_{m_l}(t,s)w_k(x)w_j(y)f(s,y)v(t,x)\gamma(y,s)\gamma(x,t)dydsdtdx \tag{4.9}$$

where the second identity is the weak convergence of the m-approximate solutions $E^+_m(f)(x,t) = u^m(x,t)$ to the weak solution $u^+(x,t)$ in $L^2([0,T];H^1(\Sigma))$ and the third identity follows from equation 4.5. We have therefore shown that we may write the kernel of the advanced Green operator as

$$G^+(t,x;s,y) = \lim_{l\to\infty}\sum_{j,k} F^{kj}_{m_l}(t,s)w_k(x)w_j(y)\gamma(y,s)\gamma(x,t). \tag{4.10}$$

Note that in terms of the above representation the fact that the formal adjoint of E^+ is E^- is reflected in the property $F(s,t) = F^T(t,s)$ for the Green matrix of a symmetric self-adjoint system of 2nd order ODEs [17].

Acknowledgments
The authors would like to acknowledge the helpful input from the referees of this paper.
YSS would like to acknowledge the financial YSS would like to acknowledge the financial support given by the Riemann fellowship.
JAV would like to acknowledge financial support by the UK STFC and the Austrian Science Fund FWF.

References

[1] C. Bär, N. Ginoux, F. Pfäffle *Wave Equations on Lorentzian Manifolds and Quantization* ESI (2007).
[2] J. Dereziński, D. Siemssen *Rev. Math. Phys.* (2017)
[3] J. Dereziński, D. Siemssen ArXiv.1709.03911 (2017)
[4] J. Dereziński, C. Gérard , *Mathematcis of Quantization and Quantum Fields*, Cambridge University Press, (2013)
[5] C. Sämann *Annal. Henri. Poincare* **17** 6 (2016)
[6] C. J. S. Clarke *Class. Quantum Grav.* **15** 975 (1998)
[7] L. C. Evans *Partial Differential Equations* American Mathematical Society, (2002).
[8] B. S. Kay, U. Lupo *Class. Quant. Grav.* **33** 215001 (2016)
[9] J. Hunter *Lecture Notes* `https://www.math.ucdavis.edu/~hunter/pdes/pde_notes.pdf`
[10] Y. Sanchez Sanchez, J.A Vickers *Class. Quant. Grav.* **33** 205002 (2016)
[11] S. W. Hawking and G. F. R. Ellis, *The large scale structure of space-time*, Cambridge University Press, (1974)
[12] P. T. Chruściel, J. D. E. Grant *Class. Quantum Grav.* **29** 145001 (2012)
[13] M. Kunzinger, R. Steinbauer, M. Stojkovic, J. Vickers *Gen. Relativ. Gravit.* **46** 1738 (2014)
[14] P. T. Chruściel *Journal of Fixed Point Theory and Applications* **4**, 325 (2013)
[15] J. P. Wilson *Class. Quantum Grav.* **17** 3199 (2000)
[16] H. Smith *Anal. PDE* **7** 1137 (2014)
[17] T. Çağri Şişman and B. Tekin *Journ. Phys. A* **43** 12 (2010)

IOP Conf. Series: Journal of Physics: Conf. Series **968** (2018) 012012 doi:10.1088/1742-6596/968/1/012012

On the proof of the C^0-inextendibility of the Schwarzschild spacetime

Jan Sbierski

Department for Pure Mathematics and Mathematical Statistics, University of Cambridge, Wilberforce Road, Cambridge, CB3 0WA, United Kingdom

E-mail: jjs48@cam.ac.uk

Abstract. This article presents a streamlined version of the author's original proof of the C^0-inextendibility of the maximal analytic Schwarzschild spacetime. Firstly, we deviate from the original proof by using the result, recently established in collaboration with Galloway and Ling, that given a C^0-extension of a globally hyperbolic spacetime, one can find a timelike geodesic that leaves this spacetime. This result much simplifies the proof of the inextendibility through the exterior region of the Schwarzschild spacetime. Secondly, we give a more flexible and shorter argument for the inextendibility through the interior region. Furthermore, we present a small new structural result for the boundary of a globally hyperbolic spacetime within a C^0-extension which serves as a new and simpler starting point for the proof.

1. Introduction

This note presents a streamlined proof of the following

Theorem 1. *The maximal analytic extension* $(M_{\max}, g_{\max})$ *of the Schwarzschild spacetime is* C^0*-inextendible.*

The original proof of this theorem given in [1] was divided into three steps: Firstly, one assumes that there is a C^0-extension of the Schwarzschild spacetime $(M_{\max}, g_{\max})$ and then shows that there is a (without loss of generality) future directed and future inextendible timelike curve γ in the Schwarzschild spacetime that can be extended to the future as a timelike curve in the C^0-extension. Secondly, one shows that the curve γ cannot lie in the exterior region of the Schwarzschild black hole and thus leave the spacetime through timelike or null infinity. Here, one uses the future one-connectedness[1] of the exterior region together with its future divergence[2]. Especially the property of future one-connectedness is quite tedious to establish. Thirdly, one shows that the timelike curve γ can neither be ultimately contained in the interior of the Schwarzschild black hole and thus leave through the curvature singularity at $r = 0$. Here, one establishes that the *spacelike diameter*[3] of a certain spherically symmetric subset 'touching' the curvature singularity, is infinite. Covering this spherically symmetric subset by a finite number

[1] Recall that a time-oriented Lorentzian manifold is called *future one-connected* iff any two timelike curves with the same endpoints are homotopic with fixed endpoints via timelike curves; see also Definition 2.13 of [1].

[2] This is the property, that for any future directed inextendible timelike curve $\gamma : [0, 1) \to M$ one has $d(\gamma(0), \gamma(s)) \to \infty$ for $s \to 1$, where d is the Lorentzian distance function.

[3] This is a geometric quantity at the C^0-level of the metric.

of charts in which the metric is suitably controlled, and which are obtained from the assumed C^0-extension, one obtains the contradiction that the spacelike diameter should be finite. For the construction of the finitely many charts one in particular exploits the spherical symmetry of the Schwarzschild spacetime.

The strategy of the proof given in this note deviates as follows from the original strategy: The first step is replaced by a result established in [2], which states that given a C^0-extension of a globally hyperbolic Lorentzian manifold (M, g) there exists a (without loss of generality) future directed and future inextendible timelike *geodesic* in M which has a future limit point on the boundary of M in the extension. Applying this result to the maximal analytic Schwarzschild spacetime, one again distinguishes the two cases of the timelike geodesic leaving through timelike/null infinity or through the curvature singularity at $r = 0$. The first case is now easily ruled out using the completeness of geodesics approaching timelike or null infinity – without the need of establishing the future one-connectedness of the exterior. For the second case we present a new, *local* version of the obstruction to C^0-extensions through $r = 0$, which does not rely on the infinitude of the spacelike diameter of the *global* spherically symmetric set constructed in [1] (which has to be covered by multiple charts), but transfers the idea locally to a single chart. This new argument is more flexible and can be more easily adapted to other situations; this part of the argument does not rely on spherical symmetry. We should also point out that the proof overall still requires the future one-connectedness of the interior.

For the sake of brevity this note is not self-contained. We will often refer the reader to definitions and results from [1]. We begin by presenting a small new structural result for C^0-extensions which is convenient to have at one's disposal.

2. A structural result for C^0-extensions

This section establishes an easy yet fundamental result on the structure of C^0-extensions. We implicitly assume that all manifolds are connected. We then recall from Section 2.3 of [1] that a *C^0-extension* of a smooth Lorentzian manifold (M, g) consists of a Lorentzian manifold $(\tilde{M}, \tilde{g})$ of the same dimension as M, $\tilde{g} \in C^0$, together with an isometric embedding $\iota : M \hookrightarrow \tilde{M}$, such that $\iota(M)$ is a proper subset of $\tilde{M}$. If no such C^0-extension exists, we say that (M, g) is *C^0-inextendible*. Finally, we recall that it is convenient for our purposes to define a *timelike curve* to be a piecewise smooth curve such that, where defined, the tangent vector is everywhere timelike and at the points of discontinuity, the left and right sided tangent vectors are timelike and lie in the same connectedness component of the timelike cone. We recall from Proposition 2.6 of [1] that, with this definition of timelike curve, the timelike future and past of a point in a time-oriented Lorentzian manifold remains open even for merely continuous metrics.

In the following we define the future (past) boundary of a time-oriented Lorentzian manifold with respect to a C^0-extension (see also Definition 2.1 in [3]).

Definition 1. *Let (M, g) be a time-oriented Lorentzian manifold with a continuous metric and $\iota : M \hookrightarrow \tilde{M}$ a C^0-extension of M. The* future boundary *of M is the set $\partial^+\iota(M)$ consisting of all points $p \in \tilde{M}$ such that there exists a timelike curve $\tilde{\gamma} : [-1, 0] \to \tilde{M}$ such that $\mathrm{Im}(\tilde{\gamma}|_{[-1,0)}) \subseteq \iota(M)$, $\tilde{\gamma}(0) = p \in \partial\iota(M)$, and $\iota^{-1} \circ \tilde{\gamma}|_{[-1,0)}$ is future directed in M.*

Clearly we have $\partial^+\iota(M) \subseteq \partial\iota(M)$. The past boundary $\partial^-\iota(M)$ is defined analogously. Moreover, it follows from Lemma 2.17 of [1] that $\partial^+\iota(M) \cup \partial^-\iota(M) \neq \emptyset$.

Figure 1 below gives a first idea of possible boundary structures. In the left Penrose-like diagram the segment from q to r, closed at q, open at r, consists of future boundary points. The segment from p to r, closed at p, open at r, consists of past boundary points. The boundary point r lies neither in the future nor the past boundary. And finally all points of the open segment from p to q lie in the future as well as in the past boundary.

The right diagram shows a C^0-extension of a globally hyperbolic Lorentzian manifold. Again, the future and past boundaries are not disjoint. Moreover, the future (past) boundary is not achronal in the extension, nor in any small neighbourhood of the future boundary point s.

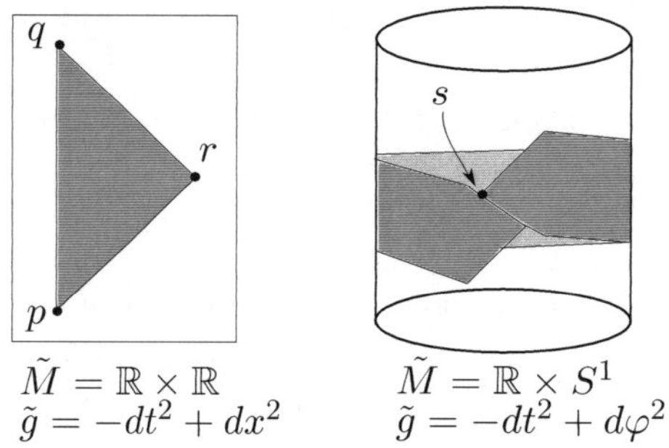

Figure 1. Possible structures of the future and past boundaries

However, we have the following result for globally hyperbolic (M, g).

Proposition 1. *Let $\iota : M \hookrightarrow \tilde{M}$ be a C^0-extension of a globally hyperbolic Lorentzian manifold (M, g) and let $p \in \partial^+\iota(M)$. For every $\delta > 0$ there exists a chart $\tilde{\varphi} : \tilde{U} \to (-\varepsilon_0, \varepsilon_0) \times (-\varepsilon_1, \varepsilon_1)^d$, $\varepsilon_0, \varepsilon_1 > 0$ with the following properties*

i) $p \in \tilde{U}$ *and* $\tilde{\varphi}(p) = (0, \ldots, 0)$

ii) $|\tilde{g}_{\mu\nu} - m_{\mu\nu}| < \delta$, *where* $m_{\mu\nu} = \mathrm{diag}(-1, 1, \ldots, 1)$

iii) There exists a Lipschitz continuous function $f : (-\varepsilon_1, \varepsilon_1)^d \to (-\varepsilon_0, \varepsilon_0)$ with the following property:

$$\{(x_0, \underline{x}) \in (-\varepsilon_0, \varepsilon_0) \times (-\varepsilon_1, \varepsilon_1)^d \mid x_0 < f(\underline{x})\} \subseteq \tilde{\varphi}(\iota(M) \cap \tilde{U}) \tag{2.1}$$

and

$$\{(x_0, \underline{x}) \in (-\varepsilon_0, \varepsilon_0) \times (-\varepsilon_1, \varepsilon_1)^d \mid x_0 = f(\underline{x})\} \subseteq \tilde{\varphi}(\partial^+\iota(M) \cap \tilde{U}) . \tag{2.2}$$

Moreover, the set on the left hand side of (2.2), i.e. the graph of f, is achronal in $(-\varepsilon_0, \varepsilon_0) \times (-\varepsilon_1, \varepsilon_1)^d$.

Remark 1. *(i) Note that the achronality of the graph of f in particular entails that any past directed timelike curve starting below the graph of f remains below the graph of f (and thus in $\tilde{\varphi}(\iota(M) \cap \tilde{U})$). This will be used often.*

(ii) As shown by the proof, given a Cauchy hypersurface Σ of M it can also be arranged that we have in fact

$$\{(x_0, \underline{x}) \in (-\varepsilon_0, \varepsilon_0) \times (-\varepsilon_1, \varepsilon_1)^d \mid x_0 < f(\underline{x})\} \subseteq \tilde{\varphi}(\iota(I^+(\Sigma, M)) \cap \tilde{U}) ,$$

Proof. Since $p \in \partial^+\iota(M)$, there exists a timelike curve $\tilde{\gamma} : [-1, 0] \to \tilde{M}$ with $\mathrm{Im}(\tilde{\gamma}|_{[-1,0)}) \subseteq \iota(M)$ and $\tilde{\gamma}(0) = p \in \partial(\iota(M))$. After a possible reparametrisation of $\tilde{\gamma}$ there exists a chart $\tilde{\varphi} : \tilde{U} \to (-\varepsilon_0, \varepsilon_0) \times (-\varepsilon_1, \varepsilon_1)^d$ such that (see also [1], Lemma 2.4)

i) $(\tilde{\varphi} \circ \tilde{\gamma})(s) = (s, 0, \ldots, 0)$ for $s \in (-\varepsilon_0, 0]$

ii) $|\tilde{g}_{\mu\nu} - m_{\mu\nu}| < \delta$

In particular $\delta > 0$ can be chosen so small that $-\frac{1}{2}dx_0^2 + dx_1^2 + \ldots + dx_d^2 \prec \tilde{g}$, which means that the light cones of the first Lorentzian metric are contained in the light cones of $\tilde{g}$.

Let $\Sigma \subseteq M$ be a Cauchy hypersurface of M and let $\gamma := \iota^{-1} \circ \tilde{\gamma}|_{[-1,0)}$. For $s_0 < 0$ close enough to 0 we have $\gamma(s_0) \in I^+(\Sigma, M)$. Since $I^+(\Sigma, M)$ is an open set, we can choose $\varepsilon_0, \varepsilon_1$

smaller, if necessary, with $-\varepsilon_0 < s_0 < 0$, such that $(-\varepsilon_0, s_0] \times (-\varepsilon_1, \varepsilon_1)^d \subseteq \tilde{\varphi}\big(\iota(I^+(\Sigma, M)) \cap \tilde{U}\big)$. Furthermore, we can assume that $\varepsilon_1 < \frac{1}{2}\varepsilon_0$. Together with our choice of δ this implies in particular that for any point $\underline{x} \in (-\varepsilon_1, \varepsilon_1)^d$ the straight line connecting $(\frac{9}{10}\varepsilon_0, \underline{x})$ with 0 is timelike.

We now define $f : (-\varepsilon_1, \varepsilon_1)^d \to (-\varepsilon_0, \varepsilon_0]$ by

$$f(\underline{x}) = \sup\{s_0 \in (-\varepsilon_0, \varepsilon_0) \mid (s, \underline{x}) \in \tilde{\varphi}\big(\iota(I^+(\Sigma, M)) \cap \tilde{U}\big) \quad \forall s \in (-\varepsilon_0, s_0)\} \, .$$

We first show that for all $\underline{x} \in (-\varepsilon_1, \varepsilon_1)^d$ we have $f(\underline{x}) < \varepsilon_0$.

Assuming the negation, there exists an $\underline{x}_0 \in (-\varepsilon_1, \varepsilon_1)^d$ with $f(\underline{x}_0) = \varepsilon_0$. It follows that $(\frac{9}{10}\varepsilon_0, \underline{x}_0) \in \tilde{\varphi}\big(\iota(I^+(\Sigma, M)) \cap \tilde{U}\big)$. Let $\sigma = (\sigma_0, \underline{\sigma}) : [0,1] \to (-\varepsilon_0, \varepsilon_0) \times (-\varepsilon_1, \varepsilon_1)^d$ denote the past directed timelike straight line from $\sigma(0) = (\frac{9}{10}\varepsilon_0, \underline{x}_0)$ to $\sigma(1) = 0$. We claim that $\sigma|_{[0,1)}$ maps into $\tilde{\varphi}\big(\iota(I^+(\Sigma, M)) \cap \tilde{U}\big)$.

To see this we (partially) foliate the plane

$$\{(t, \underline{\sigma}(s)) \in (-\varepsilon_0, \varepsilon_0) \times (-\varepsilon_1, \varepsilon_1)^d \mid t \leq \sigma_0(s), s \in (0,1]\}$$

by the (closed) straight lines ρ_τ starting at $(\tau, \underline{x}_0)$ with slope $(\frac{9}{10}\varepsilon_0, -\underline{x}_0)$ and ending on σ. It follows from the openness of $\tilde{\varphi}\big(\iota(M) \cap \tilde{U}\big)$ that there is a $-\frac{9}{10}\varepsilon_0 < \tau' < \frac{9}{10}\varepsilon_0$ such that

$$\rho_\tau \subseteq \tilde{\varphi}\big(\iota(M) \cap \tilde{U}\big) \text{ holds } \forall \tau \in (\tau', \frac{9}{10}\varepsilon_0) \, . \tag{2.3}$$

Let $\tau_0 \in [-\frac{9}{10}\varepsilon_0, \frac{9}{10}\varepsilon_0)$ be the infimum over all such τ' with the property (2.3). We first remark that for $\tau \in (\tau_0, \frac{9}{10}\varepsilon_0)$ we have $\rho_\tau \subseteq \tilde{\varphi}\big(\iota(I^+(M, \Sigma)) \cap \tilde{U}\big)$. Assuming that $\tau_0 > -\frac{9}{10}\varepsilon_0$, there exists a point q on ρ_{τ_0} with $q \notin \tilde{\varphi}\big(\iota(M) \cap \tilde{U}\big)$. This however allows us to construct a past directed timelike curve in M which is past inextendible and completely contained in $I^+(\Sigma, M)$, see Figure 2. This is a contradiction to Σ being a Cauchy hypersurface of M, and thus we obtain $\tau_0 = -\frac{9}{10}\varepsilon_0$. This shows that $\sigma|_{[0,1)}$ maps into $\tilde{\varphi}\big(\iota(I^+(\Sigma, M)) \cap \tilde{U}\big)$.

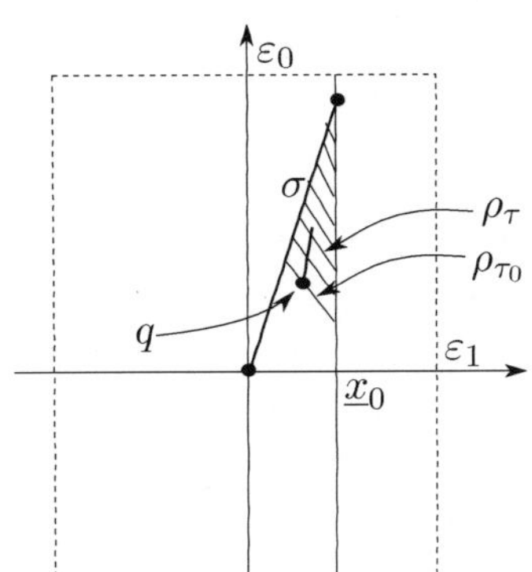

Figure 2. The continuity argument

Again, the curve $\sigma|_{[0,1)}$ corresponds to a past directed timelike curve in M which is past inextendible and completely contained in $I^+(\Sigma, M)$ – a contradiction. We thus obtain $f(\underline{x}) < \varepsilon_0$ for all $\underline{x} \in (-\varepsilon_1, \varepsilon_1)^d$.

IOP Conf. Series: Journal of Physics: Conf. Series **968** (2018) 012012 doi:10.1088/1742-6596/968/1/012012

The properties (2.1) and (2.2) are immediate. To show that the function f is continuous we use a similar construction as before: Let $\underline{x}_n \in (-\varepsilon_1, \varepsilon_1)^d$ be a sequence with $\underline{x}_n \to \underline{x}_\infty \in (-\varepsilon_1, \varepsilon_1)^d$. Assume that $f(\underline{x}_n) \not\to f(\underline{x}_\infty)$. Then there exists a $\mu > 0$ and a subsequence n_k such that $|f(\underline{x}_{n_k}) - f(\underline{x}_\infty)| > \mu$. After possibly extracting another subsequence we assume $f(\underline{x}_{n_k}) > f(\underline{x}_\infty) + \mu$; the case $f(\underline{x}_{n_k}) < f(\underline{x}_\infty) - \mu$ is dealt with analogously. For k big enough we can connect $(f(\underline{x}_{n_k}) - \frac{\mu}{2}, \underline{x}_{n_k})$ to $(f(\underline{x}_\infty), \underline{x}_\infty)$ by a straight line that is timelike and past directed. Again, we call this line σ and proceed from here along the same lines as before in order to show that σ (with the past endpoint $(f(\underline{x}_\infty), \underline{x}_\infty)$ deleted) corresponds to a past directed and past inextendible timelike curve in M which lies completely in $I^+(\Sigma, M)$. This is again a contradiction. Indeed, this argument, together with the uniform bounds on the metric components $\tilde{g}_{\mu\nu}$ also shows that f satisfies a Lipschitz condition.

In order to show that the graph of f is achronal in $(-\varepsilon_0, \varepsilon_0) \times (-\varepsilon_1, \varepsilon_1)^d$, we first observe that we have in fact shown

$$\{(x_0, \underline{x}) \in (\varepsilon_0, \varepsilon_0) \times (-\varepsilon_1, \varepsilon_1)^d \mid x_0 < f(\underline{x})\} \subseteq \tilde{\varphi}\big(\iota\big(I^+(\Sigma, M)\big) \cap \tilde{U}\big)\,.$$

Given now two points q, r in the graph of f with $r \in I^+(q, (-\varepsilon_0, \varepsilon_0) \times (-\varepsilon_1, \varepsilon_1)^d)$, we can move r down a bit in the x_0 coordinate to obtain a point which lies in $I^+(q, (-\varepsilon_0, \varepsilon_0) \times (-\varepsilon_1, \varepsilon_1)^d) \cap \tilde{\varphi}\big(\iota\big(I^+(\Sigma, M)\big) \cap \tilde{U}\big)$. It follows that we can again find a past directed and past inextendible timelike curve in M which lies completely in $I^+(\Sigma, M)$. This finishes the proof. □

Remark 2. *The proof would simplify drastically if one knew that one can find a neighbourhood of p in $\tilde{M}$ that is disjoint from a Cauchy hypersurface of M. In general this is however not the case as illustrated by the point s in Figure 1 and we need to resort to the construction illustrated in Figure 2 in order to ensure that the past directed timelike curve σ remains in $I^+(\Sigma, M)$ and does not intersect the Cauchy hypersurface Σ.*

3. Auxiliary results

The following theorem was proven in [2].

Theorem 2. *Let (M, g) be a globally hyperbolic Lorentzian manifold and $\iota : M \hookrightarrow \tilde{M}$ a C^0-extension. Assume that $\partial^+\iota(M) \neq \emptyset$. Then there exists a future directed timelike geodesic $\tau : [-1, 0) \to M$ that is future inextendible in M and such that $\lim_{s\to 0}(\iota \circ \tau)(s)$ exists and is contained in $\partial\iota(M)$.*

In fact, the proof in [2] together with Proposition 1 gives

Theorem 3. *Let (M, g) be a globally hyperbolic Lorentzian manifold and $\iota : M \hookrightarrow \tilde{M}$ a C^0-extension. Assume that $\partial^+\iota(M) \neq \emptyset$ and let $p \in \partial^+\iota(M)$. Let $\tilde{\varphi} : \tilde{U} \to (-\varepsilon_0, \varepsilon_0) \times (-\varepsilon_1, \varepsilon_1)^d$ be a chart around p as in Proposition 1. Then there exists a future directed timelike geodesic $\tau : [-1, 0) \to M$ that is future inextendible in M and such that $\tilde{\varphi} \circ \iota \circ \tau : [-1, 0) \to (-\varepsilon_0, \varepsilon_0) \times (-\varepsilon_1, \varepsilon_1)^d$ maps into $\{(s, \underline{x}) \in (-\varepsilon_0, \varepsilon_0) \times (-\varepsilon_1, \varepsilon_1)^d \mid s < f(\underline{x})\}$ and has an endpoint on $\{(s, \underline{x}) \in (-\varepsilon_0, \varepsilon_0) \times (-\varepsilon_1, \varepsilon_1)^d \mid s = f(\underline{x})\}$.*

In particular, Theorem 3 improves on Theorem 2 by guaranteeing that the future endpoint of $\iota \circ \tau$ is contained in the *future* boundary of M.

We recall from Section 4.1 of [1] the definition of the interior $(M_{\text{int}}, g_{\text{int}})$ of the $d+1$-dimensional Schwarzschild spacetime, where $M_{\text{int}} := \mathbb{R} \times \big(0, (2m)^{\frac{1}{d-2}}\big) \times \mathbb{S}^{d-1}$ and

$$g_{\text{int}} = -\big(1 - \frac{2m}{r^{d-2}}\big)\, dt^2 + \big(1 - \frac{2m}{r^{d-2}}\big)^{-1} dr^2 + r^2\, \mathring{\gamma}_{d-1}\,.$$

Here, $d \in \mathbb{N}_{\geq 3}$, (t, r) are the standard coordinates on the first two factors of M_{int}, $m > 0$ is a parameter, and $\mathring{\gamma}_{d-1}$ denotes the standard round metric on $\mathbb{S}^{d-1}$. We define a time-orientation on

IOP Conf. Series: Journal of Physics: Conf. Series **968** (2018) 012012 doi:10.1088/1742-6596/968/1/012012

$(M_{\text{int}}, g_{\text{int}})$ by stipulating that $-\frac{\partial}{\partial r}$ is future directed. For the definition of the $d+1$-dimensional maximal analytic Schwarzschild spacetime $(M_{\max}, g_{\max})$ we refer the reader to Section 4.1 of [1].

We need the following facts about the Schwarzschild spacetime:

Theorem 4. *Let $\tau : [-1, 0) \to M_{\max}$ be a future directed and future inextendible timelike geodesic (not necessarily affinely parametrised) in the maximal analytic Schwarzschild spacetime. Then we either have $(r \circ \tau)(s) \to 0$ for $s \to 0$ or τ is future complete.*

For the proof of this theorem see for example Proposition 36 in Chapter 13 of [4]. The next two results were proven in Section 6 of [1].

Lemma 1. *Let $0 < r_0 < (2m)^{\frac{1}{d-2}}$. For every $\varepsilon > 0$ we can find $0 < \tilde{r}_0 < r_0$ such that for any future directed timelike curve $\sigma : (-r_0, 0) \to M_{\text{int}}$, $\sigma(s) = \big(\sigma_t(s), -s, \sigma_\omega(s)\big)$, where σ_ω is the canonical projection of σ on the sphere $\mathbb{S}^{d-1}$, we have for all $-\tilde{r}_0 \leq s, s' < 0$*

$$d_{\mathbb{S}^{d-1}}\big(\sigma_\omega(s), \sigma_\omega(s')\big) < \varepsilon \qquad \text{and} \qquad |\sigma_t(s) - \sigma_t(s')| < \varepsilon\,.$$

Note that this lemma implies, in particular, that $\sigma_t(s)$ and $\sigma_\omega(s)$ converge for $s \nearrow 0$. We include its proof here because part of it will be referred to later on.

Proof. Let $\sigma : (-r_0, 0) \to M_{\text{int}}$ be a timelike curve, parametrised as above. We obtain for all $s \in (-r_0, 0)$

$$0 > g_{\text{int}}(\dot{\sigma}(s), \dot{\sigma}(s)) = -\big(1 - \frac{2m}{(-s)^{d-2}}\big)(\dot{\sigma}_t(s))^2 + \big(1 - \frac{2m}{(-s)^{d-2}}\big)^{-1} + s^2\, \mathring{\gamma}_{d-1}\big(\dot{\sigma}_\omega(s), \dot{\sigma}_\omega(s)\big)$$

and hence

$$\frac{(-s)^{d-2}}{2m - (-s)^{d-2}} > \underbrace{\frac{2m - (-s)^{d-2}}{(-s)^{d-2}}(\dot{\sigma}_t(s))^2}_{\geq 0} + \underbrace{s^2\, \mathring{\gamma}_{d-1}\big(\dot{\sigma}_\omega(s), \dot{\sigma}_\omega(s)\big)}_{\geq 0}\,.$$

It follows that

$$|\dot{\sigma}_t(s)| < \frac{(-s)^{d-2}}{2m - (-s)^{d-2}} \tag{3.1}$$

and

$$||\dot{\sigma}_\omega(s)||_{\mathbb{S}^{d-1}} < \frac{(-s)^{d/2-2}}{\big[2m - (-s)^{d-2}\big]^{1/2}}$$

holds for all $s \in (-r_0, 0)$. Since $d \geq 3$, it follows that both upper bounds are integrable on $(-r_0, 0)$. The lemma now follows from integration. □

Proposition 2. *The interior of the Schwarzschild spacetime $(M_{\text{int}}, g_{\text{int}})$ is future one-connected.*

4. The proof of the C^0-inextendibility of the Schwarzschild spacetime

Proof of Theorem 1: The proof is by contradiction, so we assume that there exists a C^0-extension $\iota : M_{\max} \hookrightarrow \tilde{M}$ of the maximal analytic Schwarzschild spacetime $(M_{\max}, g_{\max})$. We recall that Lemma 2.17 of [1] implies that $\partial^+\iota(M_{\max}) \cup \partial^-\iota(M_{\max}) \neq \emptyset$. Without loss of generality we assume that $\partial^+\iota(M_{\max}) \neq \emptyset$, otherwise we reverse the time orientation. By Theorem 3 there is a chart $\tilde{\varphi} : \tilde{U} \to (-\varepsilon_0, \varepsilon_0) \times (-\varepsilon_1, \varepsilon_1)^d$, $\varepsilon_0, \varepsilon_1 > 0$, and a future directed timelike geodesic $\tau : [-1, 0) \to M_{\max}$ that is future inextendible in $M_{\max}$ with the following properties:

i) $|\tilde{g}_{\mu\nu} - m_{\mu\nu}| < \delta_0$ (where $\delta_0 > 0$ is fixed below).

ii) There exists a Lipschitz continuous function $f : (-\varepsilon_1, \varepsilon_1)^d \to (-\varepsilon_0, \varepsilon_0)$ with the following property:

$$\{(x_0, \underline{x}) \in (-\varepsilon_0, \varepsilon_0) \times (-\varepsilon_1, \varepsilon_1)^d \mid x_0 < f(\underline{x})\} \subseteq \tilde{\varphi}\big(\iota(M_{\max}) \cap \tilde{U}\big) \tag{4.1}$$

and

$$\{(x_0, \underline{x}) \in (-\varepsilon_0, \varepsilon_0) \times (-\varepsilon_1, \varepsilon_1)^d \mid x_0 = f(\underline{x})\} \subseteq \tilde{\varphi}\big(\partial^+\iota(M_{\max}) \cap \tilde{U}\big) \,. \tag{4.2}$$

Moreover, the graph of f is achronal in $\tilde{U}$.

iii) $\tilde{\varphi} \circ \iota \circ \tau : [-1, 0) \to (-\varepsilon_0, \varepsilon_0) \times (-\varepsilon_1, \varepsilon_1)^d$ maps into $\{(x_0, \underline{x}) \in (-\varepsilon_0, \varepsilon_0) \times (-\varepsilon_1, \varepsilon_1)^d \mid x_0 < f(\underline{x})\}$ and $\lim_{s \to 0}(\tilde{\varphi} \circ \iota \circ \tau)(s) = (0, \dots, 0)$.[4]

We will also introduce the abbreviation $R_{\varepsilon_0,\varepsilon_1} := (-\varepsilon_0, \varepsilon_0) \times (-\varepsilon_1, \varepsilon_1)^d$. Let $0 < a < 1$ and let $< \cdot, \cdot >_{R^{d+1}}$ denote the Euclidean inner product on $\mathbb{R}^{d+1}$ and $| \cdot |_{\mathbb{R}^{d+1}}$ the associated norm. We introduce the following notation:

- $C_a^+ := \big\{X \in \mathbb{R}^{d+1} \mid \frac{<X, e_0>_{\mathbb{R}^{d+1}}}{|X|_{\mathbb{R}^{d+1}}} > a\big\}$
- $C_a^- := \big\{X \in \mathbb{R}^{d+1} \mid \frac{<X, e_0>_{\mathbb{R}^{d+1}}}{|X|_{\mathbb{R}^{d+1}}} < -a\big\}$
- $C_a^c := \big\{X \in \mathbb{R}^{d+1} \mid -a < \frac{<X, e_0>_{\mathbb{R}^{d+1}}}{|X|_{\mathbb{R}^{d+1}}} < a\big\}$.

Here, C_a^+ is the forward cone of vectors which form an angle of less than $\cos^{-1}(a)$ with the x_0-axis, and C_a^- is the corresponding backwards cone. In Minkowski space, the forward and backward cones of timelike vectors correspond to the value $a = \cos(\frac{\pi}{4}) = \frac{1}{\sqrt{2}}$.

Since $\frac{5}{8} < \frac{1}{\sqrt{2}} < \frac{5}{6}$, we can now choose $\delta_0 > 0$ such that in the chart $\tilde{\varphi}$ from above all vectors in $C^+_{5/6}$ are future directed timelike, all vectors in $C^-_{5/6}$ are past directed timelike, and all vectors in $C^c_{5/8}$ are spacelike. It is straightforward then to prove[5] the following estimates for $x \in R_{\varepsilon_0,\varepsilon_1}$:

$$\begin{aligned} \big(x + C^+_{5/6}\big) \cap R_{\varepsilon_0,\varepsilon_1} \subseteq I^+(x, R_{\varepsilon_0,\varepsilon_1}) \subseteq \big(x + C^+_{5/8}\big) \cap R_{\varepsilon_0,\varepsilon_1} \\ \big(x + C^-_{5/6}\big) \cap R_{\varepsilon_0,\varepsilon_1} \subseteq I^-(x, R_{\varepsilon_0,\varepsilon_1}) \subseteq \big(x + C^-_{5/8}\big) \cap R_{\varepsilon_0,\varepsilon_1} \,. \end{aligned} \tag{4.3}$$

By Theorem 4 the future directed timelike geodesic $\tau : [-1, 0) \to M_{\max}$ is either future complete (i.e., it 'leaves through the exterior') or $(r \circ \tau)(s) \to 0$ holds for $s \to 0$ (i.e., it 'leaves through the interior'). We first show that it cannot leave through the exterior.

In order to ease notation we also denote $\tilde{\varphi} \circ \iota \circ \tau$ in the following by τ. Since $\dot{\tau}(s) \in C^+_{5/8}$ for all $s \in [-1, 0)$, we have $dx_0\big(\dot{\tau}(s)\big) > 0$. It follows that we can reparametrise τ so that we can assume without loss of generality that $\tau : [-s_0, 0) \to R_{\varepsilon_0,\varepsilon_1}$ is given by $\tau(s) = \big(s, \underline{\tau}(s)\big)$, where $-s_0 \in (-\varepsilon_0, 0)$. From $\dot{\tau}(s) \in C^+_{5/8}$ for all $s \in [-s_0, 0)$, it follows that

$$\frac{5}{8} < \frac{< \dot{\tau}(s), e_0 >_{\mathbb{R}^{d+1}}}{|\dot{\tau}(s)|_{\mathbb{R}^{d+1}}} = \frac{1}{\sqrt{1 + |\dot{\underline{\tau}}(s)|_{\mathbb{R}^d}}} \,.$$

Hence, we obtain $|\dot{\underline{\tau}}(s)|_{\mathbb{R}^d} < \frac{\sqrt{39}}{5}$ for all $s \in [-s_0, 0)$. Together with the uniform bound on the metric components i), it now follows that

$$\int_{-s_0}^{0} \sqrt{-\tilde{g}\big(\dot{\tau}(s), \dot{\tau}(s)\big)}\, ds = \int_{-s_0}^{0} \sqrt{-\Big[\tilde{g}_{00} + 2\sum_{i=1}^{d} \tilde{g}_{0i}\dot{\underline{\tau}}_i(s) + \sum_{i,j=1}^{d} \tilde{g}_{ij}\dot{\underline{\tau}}_i(s)\dot{\underline{\tau}}_j(s)\Big]}\, ds \le C < \infty \,.$$

[4] This can be guaranteed after recentering the chart.
[5] See also [1], Step 1.2 of the proof of Theorem 3.1.

Thus, τ is not future complete. We are left with the possibility that τ leaves through the interior, i.e., $(r \circ \tau)(s) \to 0$ holds for $s \to 0$. In the following we will also derive a contradiction from this premiss.

We consider the curve $\tilde{\gamma} : (-\varepsilon_0, 0] \to \tilde{M}$ which is given in the chart $\tilde{\varphi}$ by $(-\varepsilon_0, 0] \ni s \mapsto (s, 0, \ldots, 0)$ and we set $\gamma := \iota^{-1} \circ \tilde{\gamma}|_{(-\varepsilon_0, 0)}$, which is a future directed and future inextendible timelike curve in $M_{\max}$. We claim that $(r \circ \gamma)(s) \to 0$ for $s \to 0$. To see this, we observe that by Proposition 2.6 of [1] $I^+(\tilde{\gamma}(s), \tilde{U})$ is an open neighbourhood of $(0, \ldots, 0)$ for every $s \in (-\varepsilon_0, 0)$. Thus, there exists a future directed timelike curve from $\tilde{\gamma}(s)$ to $(\tilde{\varphi} \circ \iota \circ \tau)([-1, 0))$ for all $s \in (-\varepsilon_0, 0)$. By the first point of Remark 1 these timelike curves lie completely in $\iota(M_{\max})$. It is easy now to conclude that γ must also leave through the interior[6]. From now on we can forget about τ – we will only work with γ now. The rest of the proof proceeds in three steps.

Step 1: We show that there exists a $\mu > 0$ such that

(i) $\iota\Big(I^+\big(\gamma(-\mu), M_{\text{int}}\big)\Big) \subseteq \tilde{\varphi}^{-1}\Big(\{(x_0, \underline{x}) \in (\varepsilon_0, \varepsilon_0) \times (-\varepsilon_1, \varepsilon_1)^d \mid x_0 < f(\underline{x})\}\Big)$

(ii) $(-\varepsilon_0, -\frac{49}{50}\varepsilon_0) \times (-\varepsilon_1, \varepsilon_1)^d \subseteq I^-\big((\tilde{\varphi} \circ \tilde{\gamma})(-\mu), (-\varepsilon_0, \varepsilon_0) \times (-\varepsilon_1, \varepsilon_1)^d\big)$.

Step 1.1 As in the proof of Proposition 1 we can without loss of generality assume that $0 < \varepsilon_1 < \frac{1}{2}\varepsilon_0$. We now choose $x^+ := (x_0^+, 0, \ldots, 0)$, $0 < x_0^+ < \varepsilon_0$, and $x^- := (x_0^-, 0, \ldots, 0)$, $-\varepsilon_0 < x_0^- < 0$ such that the closure of $(x^+ + C^-_{5/6}) \cap (x^- + C^+_{5/6})$ in $R_{\varepsilon_0,\varepsilon_1}$ is compact. Now choose $y^- := (y_0^-, 0, \ldots, 0)$ with $-\frac{1}{5}\varepsilon_0 < y_0^- < 0$ so that the closure of $C^-_{5/8} \cap (y^- + C^+_{5/8})$ in $R_{\varepsilon_0,\varepsilon_1}$ is contained in $x^+ + C^-_{6/7} \cap x^- + C^+_{6/7}$. Moreover, since we have $-\frac{1}{5}\varepsilon_0 < y_0^-$ and $0 < \varepsilon_1 < \frac{1}{2}\varepsilon_0$ it follows that

$$(-\varepsilon_0, -\frac{49}{50}\varepsilon_0) \times (-\varepsilon_1, \varepsilon_1)^d \subseteq I^-\big(y^-, R_{\varepsilon_0,\varepsilon_1}\big) . \tag{4.4}$$

We claim that for all $y_0^- < s < 0$ we have

$$\tilde{\varphi}^{-1}\Big(I^-\big((s, 0, \ldots, 0), R_{\varepsilon_0,\varepsilon_1}\big) \cap I^+\big(y^-, R_{\varepsilon_0,\varepsilon_1}\big)\Big) = \iota\Big(I^-\big(\gamma(s), M_{\text{int}}\big) \cap I^+\big(\gamma(y_0^-), M_{\text{int}}\big)\Big) . \tag{4.5}$$

The inclusion "$\subseteq$" follows from the first point of Remark 1, since if σ is a past directed timelike curve in $R_{\varepsilon_0,\varepsilon_1}$ from $(s, 0, \ldots, 0)$ to y^-, then the first point of Remark 1 states that $\tilde{\varphi}^{-1} \circ \sigma$ is contained in $\iota(M_{\text{int}})$.

To prove "$\supseteq$", let $\sigma : [y_0^-, s] \to M_{\text{int}}$ be a future directed timelike curve from $\gamma(y_0^-)$ to $\gamma(s)$. By Proposition 2, there exists a timelike homotopy $\Gamma : [0,1] \times [y_0^-, s] \to M_{\text{int}}$ with fixed endpoints between $\gamma|_{[y_0^-, s]}$ and σ. It follows that $\iota \circ \Gamma : [0,1] \times [y_0^-, s] \to \tilde{M}$ is a timelike homotopy with fixed endpoints in $\tilde{M}$. We need to show that $(\iota \circ \sigma)(\cdot) = (\iota \circ \Gamma)(1, \cdot)$ maps into $\tilde{U}$. We argue by continuity, i.e., we show that the interval $J := \{t \in [0,1] \mid \iota \circ \Gamma(t, [y_0^-, s]) \subseteq \tilde{U}\}$ is non-empty, open and closed in $[0,1]$.

Clearly, we have $0 \in J$, since $\iota \circ \Gamma(0, \cdot) = \tilde{\gamma}|_{[y_0^-, s]}$. The openness follows from the openness of $\tilde{U}$, and the closedness follows since $I^-\big(\tilde{\gamma}(s), \tilde{U}\big) \cap I^+\big(\tilde{\gamma}(y_0^-), \tilde{U}\big)$ is precompact in $\tilde{U}$, i.e., in particular its closure in $\tilde{M}$ is contained in $\tilde{U}$. This finishes the proof of (4.5).

Moreover, it is easy to see[7] that the following holds

$$I^-\big(0, R_{\varepsilon_0,\varepsilon_1}\big) \cap I^+\big(y^-, R_{\varepsilon_0,\varepsilon_1}\big) = \bigcup_{y_0^- < s < 0} \Big(I^-\big((s, 0, \ldots, 0), R_{\varepsilon_0,\varepsilon_1}\big)\Big) \cap I^+\big(y^-, R_{\varepsilon_0,\varepsilon_1}\big) .$$

[6] Indeed, through the same point of the conformal boundary as given by the Penrose conformal compactification.
[7] See also [1], Proposition 2.7

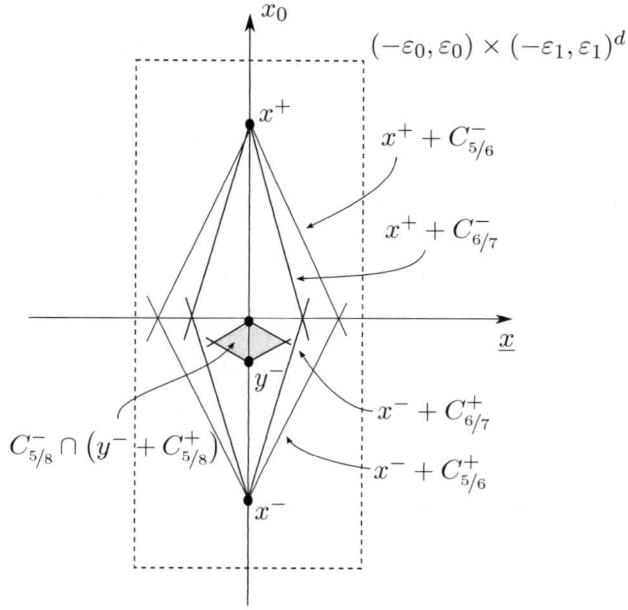

Figure 3. The set-up of Step 1.1

Together with (4.5) this now implies

$$\tilde{\varphi}^{-1}\Big(I^-\big(0, R_{\varepsilon_0,\varepsilon_1}\big) \cap I^+\big(y^-, R_{\varepsilon_0,\varepsilon_1}\big)\Big) = \iota\Big(\Big(\bigcup_{y_0^- < s < 0} I^-\big(\gamma(s), M_{\mathrm{int}}\big)\Big) \cap I^+\big(\gamma(y_0^-), M_{\mathrm{int}}\big)\Big) . \quad (4.6)$$

For the next step we recall from [1], Definition 2.14, that for a Lorentzian manifold (M, g) two sets $A, B \subseteq M$ are called *timelike separated by a set* $K \subseteq M$, iff every timelike curve connecting A and B intersects K.

Step 1.2 In this step we switch back to the manifold $(M_{\mathrm{int}}, g_{\mathrm{int}})$. Choose a y_0^+ with $y_0^- < y_0^+ < 0$ and define

$$K := \Big[\Big(\bigcup_{y_0^- < s < 0} I^-\big(\gamma(s), M_{\mathrm{int}}\big)\Big) \cap I^+\big(\gamma(y_0^-), M_{\mathrm{int}}\big)\Big] \setminus I^+\big(\gamma(y_0^+), M_{\mathrm{int}}\big) . \quad (4.7)$$

Step 1.2.1 We show that the set K timelike separates the set $\gamma\big((y_0^+, 0)\big)$ from $I^-\big(\gamma(y_0^-), M_{\mathrm{int}}\big)$.

So let $\sigma : [0, 1] \to M_{\mathrm{int}}$ be a past directed timelike curve with $\sigma(0) \in \gamma\big((y_0^+, 0)\big)$ and $\sigma(1) \in I^-\big(\gamma(y_0^-), M_{\mathrm{int}}\big)$.

We claim that there exists a $\Delta_- \in (0, 1)$ such that

$$\sigma^{-1}\big[I^+\big(\gamma(y_0^-), M_{\mathrm{int}}\big)\big] = [0, \Delta_-) .$$

This is seen as follows: To begin with, it is clear that $0 \in \sigma^{-1}\big[I^+\big(\gamma(y_0^-), M_{\mathrm{int}}\big)\big]$. Moreover, by the continuity of σ and the openness of $I^+\big(\gamma(y_0^-), M_{\mathrm{int}}\big)$, we know that $\sigma^{-1}\big[I^+\big(\gamma(y_0^-), M_{\mathrm{int}}\big)\big]$ is open in $[0, 1]$. Moreover, since σ is a past directed timelike curve, it follows that if

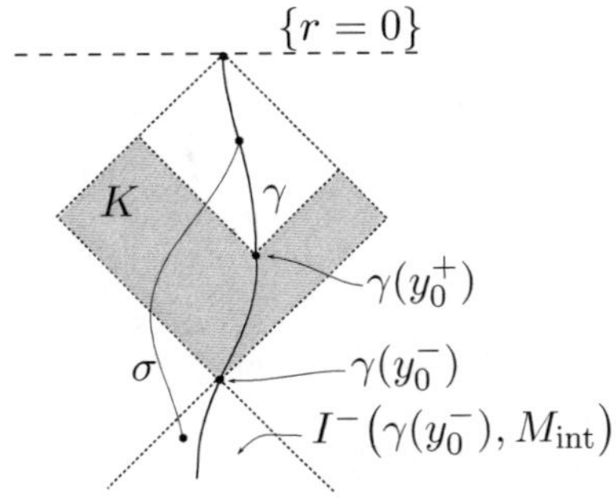

Figure 4. For the proof of Step 1.2.1

$s_0 \in \sigma^{-1}\big[I^+\big(\gamma(y_0^-), M_{\text{int}}\big)\big]$, then we also have $[0, s_0] \in \sigma^{-1}\big[I^+\big(\gamma(y_0^-), M_{\text{int}}\big)\big]$. And finally, since $(M_{\text{int}}, g_{\text{int}})$ satisfies the chronology condition, there are no closed timelike curves in M_{int}, and hence $I^-\big(\gamma(y_0^-), M_{\text{int}}\big)$ is disjoint from $I^+\big(\gamma(y_0^-), M_{\text{int}}\big)$. This implies that $\Delta_- < 1$.

In the same way we deduce that there exists a $\Delta_+ \in (0, 1)$ such that

$$\sigma^{-1}\big[I^+\big(\gamma(y_0^+), M_{\text{int}}\big)\big] = [0, \Delta_+) \, .$$

In the following we show that $\Delta_+ < \Delta_-$.

Since $(M_{\text{int}}, g_{\text{int}})$ is globally hyperbolic and g_{int} is smooth (!), we have $\overline{I^+\big(\gamma(y_0^+), M_{\text{int}}\big)} = J^+\big(\gamma(y_0^+), M_{\text{int}}\big)$.[8] Together with $\gamma(y_0^+) \in I^+\big(\gamma(y_0^-), M_{\text{int}}\big)$, we now obtain[9]

$$\overline{I^+\big(\gamma(y_0^+), M_{\text{int}}\big)} = J^+\big(\gamma(y_0^+), M_{\text{int}}\big) \subseteq J^+\Big(I^+\big(\gamma(y_0^-), M_{\text{int}}\big), M_{\text{int}}\Big) = I^+\big(\gamma(y_0^-), M_{\text{int}}\big) \, .$$

Hence, we have $\sigma(\Delta_+) \in I^+\big(\gamma(y_0^-), M_{\text{int}}\big)$, from which it follows that $\Delta_+ < \Delta_-$.

Choosing $s_0 \in (\Delta_+, \Delta_-)$, it follows that $\sigma(s_0) \in I^+\big(\gamma(y_0^-), M_{\text{int}}\big) \setminus I^+\big(\gamma(y_0^+), M_{\text{int}}\big)$. Moreover, it is clear that $\sigma(s_0) \in \bigcup_{-\varepsilon < s < 0} I^-\big(\gamma(s), M_{\text{int}}\big)$, which concludes Step 1.2.1.

Step 1.2.2 We show that $\overline{K}$ is compact.

We claim that for any $s_0 \in (y_0^+, 0)$ there exists a $\delta > 0$ and a neighbourhood V of $\text{id} \in SO(d)$ such that

$$\big(\gamma_t(s) - \delta, \gamma_t(s) + \delta\big) \times \big\{\gamma_r(s)\big\} \times \big\{f \cdot \gamma_\omega(s) \,|\, f \in V\big\} \subseteq I^+\big(\gamma(y_0^+), M_{\text{int}}\big)$$

holds for all $s \in (s_0, 0)$.

In order to prove this claim, we first note that since γ is timelike and g_{int} is continuous, there exists a $\mu > 0$ such that

$$g_{\text{int}}\big(\dot\gamma(s), \dot\gamma(s)\big) < -\mu \tag{4.8}$$

holds for all $s \in [y_0^+, s_0]$.

Let now $\lambda \in C^\infty\big([y_0^+, s_0], \mathbb{R}\big)$ and $h \in C^\infty\big([y_0^+, s_0], SO(d) \subseteq \text{Mat}(d \times d, \mathbb{R})\big)$, and define $\sigma : [y_0^+, s_0] \to M_{\text{int}}$ by

$$\sigma(s) := \Big(\gamma_t(s) + \lambda(s), \gamma_r(s), h(s)\big(\gamma_\omega(s)\big)\Big) \, .$$

[8] This follows from 6. Lemma and 22. Lemma of Chapter 14 of [4]. Here $J^+\big(\gamma(y_0^+), M_{\text{int}}\big)$ denotes the causal future of $\gamma(y_0^+)$ in M_{int}, see also Chapter 14 of [4].

[9] See 1. Corollary in Chapter 14 of [4].

We compute

$$g_{\text{int}}(\dot\sigma,\dot\sigma) = -\Big(1-\frac{2m}{\big(\gamma_r(s)\big)^{d-2}}\Big)\big(\dot\gamma_t(s)+\dot\lambda(s)\big)^2 + \Big(1-\frac{2m}{\big(\gamma_r(s)\big)^{d-2}}\Big)^{-1}\big(\dot\gamma_r(s)\big)^2 + \big(\gamma_r(s)\big)^2\,\big|\big|\dot h(s)\gamma_\omega(s)+h(s)\dot\gamma_\omega(s)\big|\big|^2_{\mathbb{R}^d}\,, \tag{4.9}$$

where $||\cdot||_{\mathbb{R}^d}$ denotes the Euclidean norm on $\mathbb{R}^d$, and we think of γ_ω as mapping into $\mathbb{S}^{d-1}\subseteq\mathbb{R}^d$. Since $\gamma_r(s)$ is bounded away from 0 for $s\in[y_0^+,s_0]$, we can infer from (4.8) and (4.9) that there exists an $\eta>0$ such that whenever

$$||\dot\lambda||_{L^\infty\left([y_0^+,s_0]\right)} + ||\dot h||_{L^\infty\left([y_0^+,s_0]\right)} < \eta \tag{4.10}$$

holds, the curve $\sigma:[y_0^+,s_0]\to M_{\text{int}}$ is timelike[10]. This in turn implies the existence of a $\delta>0$ and a neighbourhood V of $\text{id}\in SO(d)$ such that for every $\lambda_{s_0}\in(-\delta,\delta)$ and for every $h_{s_0}\in V$ there are smooth functions

$$\lambda\in C^\infty\big([y_0^+,s_0],\mathbb{R}\big) \quad \text{with } \lambda(y_0^+)=0 \text{ and } \lambda(s_0)=\lambda_{s_0}$$

and

$$h\in C^\infty\big([y_0^+,s_0],SO(d)\subseteq \text{Mat}(d\times d,\mathbb{R})\big) \quad \text{with } h(y_0^+)=\text{id} \text{ and } h(s_0)=h_{s_0}$$

such that moreover (4.10) is satisfied. The claim now follows from concatenating σ with the timelike[11] curve $\tau:[s_0,0)\to M_{\text{int}}$ given by

$$\tau(s)=\Big(\gamma_t(s)+\lambda_{s_0},\gamma_r(s),h_{s_0}\big(\gamma_\omega(s)\big)\Big)\,.$$

We now fix $s_0\in(y_0^+,0)$ and obtain $\delta>0$ and a neighbourhood $V\subseteq SO(d)$ of $\text{id}\in SO(d)$ as in the claim. It follows from Lemma 1 that there exists $s_1\in(s_0,0)$ (close to 0) such that

$$I^-\big(\gamma(s),M_{\text{int}}\big)\cap\Big\{r\le\gamma_r(s_1)\Big\}\subseteq\bigcup_{s_0\le s'<0}\Big[\big(\gamma_t(s')-\delta,\gamma_t(s')+\delta\big)\times\big\{\gamma_r(s')\big\}\times\big\{f\cdot\gamma_\omega(s')\,|\,f\in V\big\}\Big] \subseteq I^+\big(\gamma(y_0^+),M_{\text{int}}\big)$$

holds for all $s\in(-\varepsilon_0,0)$. This implies

$$K\subseteq\mathbb{R}\times\big(\gamma_r(s_1),\gamma_r(y_0^-)\big)\times\mathbb{S}^{d-1}\,.$$

Moreover, the bound (3.1) implies that there are $t_0,t_1\in\mathbb{R}$ such that

$$I^+\big(\gamma(y_0^-),M_{\text{int}}\big)\subseteq(t_0,t_1)\times\big(0,\gamma_r(y_0^-)\big)\times\mathbb{S}^{d-1}\,.$$

It follows that

$$K\subseteq(t_0,t_1)\times\big(\gamma_r(s_1),\gamma_r(y_0^-)\big)\times\mathbb{S}^{d-1}\,,$$

which implies that $\overline{K}$ is compact.

Step 1.3 First note that it follows from the definition of K, (4.7), that $K\subseteq\Big(\bigcup_{-\varepsilon_0<s<0}I^-\big(\gamma(s),M_{\text{int}}\big)\Big)\cap I^+\big(\gamma(y_0^-),M_{\text{int}}\big)$, and thus, together with (4.6), we obtain

$$\overline{\tilde\varphi\big(\iota(K)\big)}\subseteq\overline{I^-\big(0,R_{\varepsilon_0,\varepsilon_1}\big)\cap I^+\big(y^-,R_{\varepsilon_0,\varepsilon_1}\big)}\,.$$

[10] One can for example define $||\dot h||_{L^\infty\left([y_0^+,s_0]\right)} := \sup_{s\in[y_0^+,s_0]} ||\dot h(s)||_{\mathbb{R}^{d\times d}}$.

[11] Recall that the Schwarzschild metric $g_{\text{int}}=-(1-\frac{2m}{r^{d-2}})\,dt^2+(1-\frac{2m}{r^{d-2}})^{-1}\,dr^2+r^2\mathring{\gamma}_{d-1}$ is spherically symmetric and invariant under translations in t.

IOP Conf. Series: Journal of Physics: Conf. Series **968** (2018) 012012 doi:10.1088/1742-6596/968/1/012012

By the choice of $y^- \in R_{\varepsilon_0,\varepsilon_1}$ in Step 1.1, together with (4.3), it follows that

$$\overline{\tilde{\varphi}(\iota(K))} \subseteq \left(x^+ + C^-_{6/7}\right) \cap \left(x^- + C^+_{6/7}\right) . \tag{4.11}$$

Moreover, the continuity of $\tilde{\varphi}$ and ι implies[12]

$$\tilde{\varphi}\big(\iota(\overline{K})\big) \subseteq \overline{\tilde{\varphi}\big(\iota(K)\big)} . \tag{4.12}$$

Hence, from (4.12) and (4.11), it follows that

$$W := (\tilde{\varphi} \circ \iota)^{-1}\Big(\big[x^+ + C^-_{6/7}\big] \cap \big[x^- + C^+_{6/7}\big]\Big) \subseteq M_{\text{int}}$$

is an open neighbourhood of $\overline{K} \subseteq M_{\text{int}}$.

Step 1.4 We show that there exists a $\mu > 0$ such that $I^+\big(\gamma(-\mu), M_{\text{int}}\big)$ is timelike separated from $\gamma(x_0^-)$ by W.

We consider $M_{\text{int}} = \mathbb{R} \times \big(0, (2m)^{\frac{1}{d-2}}\big) \times \mathbb{S}^{d-1}$ with the metric $d_{M_{\text{int}}} : M_{\text{int}} \times M_{\text{int}} \to [0, \infty)$ given by

$$d_{M_{\text{int}}}\Big((t_1, r_1, \omega_1), (t_2, r_2, \omega_2)\Big) := |t_1 - t_2| + |r_1 - r_2| + d_{\mathbb{S}^{d-1}}(\omega_1, \omega_2) ,$$

where $(t_i, r_i, \omega_i) \in M_{\text{int}}$ for $i = 1, 2$. Since

$$M_{\text{int}} \ni (t, r, \omega) \mapsto d_{M_{\text{int}}}\big((t, r, \omega), M_{\text{int}} \setminus W\big) = \inf_{(t',r',\omega') \in M_{\text{int}} \setminus W} d_{M_{\text{int}}}\big((t, r, \omega), (t', r', \omega')\big)$$

is continuous, and $\overline{K}$ is compact and disjoint from the closed set $M_{\text{int}} \setminus W$, we infer that $d_{M_{\text{int}}}(\cdot, M_{\text{int}} \setminus W)$ must attain its minimum on $\overline{K}$, which is moreover strictly positive. It follows that there exists a $\delta > 0$ such that

$$\overline{K}_\delta := \Big\{(t, r, \omega) \in M_{\text{int}} \,\big|\, d_{M_{\text{int}}}\big((t, r, \omega), \overline{K}\big) < \delta\Big\} \subseteq W .$$

Moreover, by choosing δ slightly smaller if necessary, we can also assume that

$$B_\delta\big(\gamma(x_0^-)\big) \subseteq I^-\big(\gamma(y_0^-), M_{\text{int}}\big) . \tag{4.13}$$

Step 1.4.1 We define a metric $d_{SO(d)} : SO(d) \times SO(d) \to [0, \infty)$ on $SO(d)$ by

$$d_{SO(d)}(f, h) := \sup_{\omega \in \mathbb{S}^{d-1}} d_{\mathbb{S}^{d-1}}\big(f(\omega), h(\omega)\big) ,$$

where $f, h \in SO(d)$, and denote with $B_\eta(\text{id}) \subseteq SO(d)$ the ball of radius $\eta > 0$, centred at id, with respect to this metric. Moreover, it is easy to see that

$$\text{for } \omega_0, \omega_1 \in \mathbb{S}^{d-1} \text{ with } d_{\mathbb{S}^{d-1}}(\omega_0, \omega_1) < \eta, \text{ there exists an } h \in B_\eta(\text{id}) \text{ with } h(\omega_0) = \omega_1 . \tag{4.14}$$

In particular, h can be defined as a rotation purely in the plane $\text{span}\{\omega_0, \omega_1\} \subseteq \mathbb{R}^d$.

Continuing the proof of Step 1.4, Lemma 1 implies that there exists a $\mu \in (0, -y_0^+)$ such that for all $(t_0, r_0, \omega_0) \in I^+\big(\gamma(-\mu), M_{\text{int}}\big)$, we have

$$|t_0 - \gamma_t(-\mu)| < \frac{\delta}{2} \qquad \text{and} \qquad d_{\mathbb{S}^{d-1}}\big(\omega_0, \gamma_\omega(-\mu)\big) < \frac{\delta}{2} . \tag{4.15}$$

[12] Indeed, since $\overline{K} \subseteq M_{\text{int}}$ is compact, we actually have equality.

In the following we will show that $I^+(\gamma(-\mu), M_{\text{int}})$ is timelike separated from $\gamma(x_0^-)$ even by $\overline{K}_\delta$ (which, of course, implies Step 1.4).

So let $\sigma : [0,1] \to M_{\text{int}}$ be a past directed timelike curve with $\sigma(0) \in I^+(\gamma(-\mu), M_{\text{int}})$ and $\sigma(1) = \gamma(x_0^-)$. Also let $s_0 \in (-\mu, 0)$ be such that $\gamma_r(s_0) = \sigma_r(0)$. By (4.15) we have

$$|\sigma_t(0) - \gamma_t(s_0)| < \delta \qquad \text{and} \qquad d_{\mathbb{S}^{d-1}}\big(\sigma_\omega(0), \gamma_\omega(s_0)\big) < \delta\,.$$

Thus, by (4.14) there exists an $h \in B_\delta(\text{id}) \subseteq SO(d)$ such that $h\big(\sigma_\omega(0)\big) = \gamma_\omega(s_0)$. It now follows that the curve $\hat{\sigma} : [0,1] \to M_{\text{int}}$, given by

$$\hat{\sigma}(s) = \Big(\sigma_t(s) + [\gamma_t(s_0) - \sigma_t(0)], \sigma_r(s), h\big(\sigma_\omega(s)\big)\Big)\,,$$

is past directed timelike with $\hat{\sigma}(0) = \gamma(s_0)$ and, using the fact that $h \in B_\delta(\text{id})$ together with (4.13), $\hat{\sigma}(1) \in I^-(\gamma(y_0^-), M_{\text{int}})$. By Step 1.2.1, there exists an $\hat{s} \in [0,1]$ with $\hat{\sigma}(\hat{s}) \in K$. It now follows that $\sigma(\hat{s}) \in \overline{K}_\delta$, which concludes Step 1.4.

We now finish the proof of Step 1. We first show that $\iota\big(I^+(\gamma(-\mu), M_{\text{int}})\big) \subseteq \tilde{\varphi}^{-1}\Big(\big[x^+ + C^-_{6/7}\big] \cap \big[x^- + C^+_{6/7}\big]\Big)$.

The proof is by contradiction. So let $\sigma : [0,1] \to M_{\text{int}}$ be a future directed timelike curve with $\sigma(0) = \gamma(-\mu)$ and assume that there exists $\tilde{s} \in [0,1]$ such that $(\tilde{\varphi} \circ \iota \circ \sigma)(\tilde{s}) \notin \big[x^+ + C^-_{6/7}\big] \cap \big[x^- + C^+_{6/7}\big]$. Let

$$s_0 := \sup\big\{s' \in [0,1] \,|\, (\tilde{\varphi} \circ \iota \circ \sigma)(s) \in \big[x^+ + C^-_{6/7}\big] \cap \big[x^- + C^+_{6/7}\big] \text{ for all } s \in [0, s')\big\}\,.$$

Clearly, we have $0 < s_0 \leq 1$ and from our assumption it follows that $(\tilde{\varphi} \circ \iota \circ \sigma)(s_0) \in \partial\big(\big[x^+ + C^-_{6/7}\big] \cap \big[x^- + C^+_{6/7}\big]\big)$. Since all vectors in $C^-_{5/6}$ are past directed timelike, we can find a past directed timelike curve $\tau : [0,1] \to R_{\varepsilon_0,\varepsilon_1}$ with $\tau(0) = (\tilde{\varphi} \circ \iota \circ \sigma)(s_0)$ and $\tau(1) = x^-$, which does not intersect $\big[x^+ + C^-_{6/7}\big] \cap \big[x^- + C^+_{6/7}\big]$. For example, this curve can be chosen to lie in $\partial\Big(\big[x^+ + C^-_{6/7}\big] \cap \big[x^- + C^+_{6/7}\big]\Big)$, see Figure 5. The first point of Remark 1 implies that τ maps in $\tilde{\varphi}\big(\iota(M_{\text{int}}) \cap \tilde{U}\big)$. Hence, $(\tilde{\varphi} \circ \iota)^{-1} \circ \tau$ is a past directed timelike curve in M_{int} with $\big((\tilde{\varphi} \circ \iota)^{-1} \circ \tau\big)(0) = \sigma(s_0) \in I^+(\gamma(-\mu), M_{\text{int}})$ and $\big((\tilde{\varphi} \circ \iota)^{-1} \circ \tau\big)(1) = \gamma(x_0^-)$, which does not intersect $W = (\tilde{\varphi} \circ \iota)^{-1}\Big(\big[x^+ + C^-_{6/7}\big] \cap \big[x^- + C^+_{6/7}\big]\Big)$. This, however, is a contradiction to Step 1.4. Hence, we have shown that $\iota\big(I^+(\gamma(-\mu), M_{\text{int}})\big) \subseteq \tilde{\varphi}^{-1}\Big(\big[x^+ + C^-_{6/7}\big] \cap \big[x^- + C^+_{6/7}\big]\Big)$, which in particular implies the first point of Step 1. The second point follows from (4.4) together with $y_0^- < y_0^+ < -\mu < 0$. This concludes the proof of Step 1.

Step 2: We show that Step 1 implies the following: There exists a constant $0 < C_d < \infty$ such that for any Cauchy hypersurface Σ of M_{int} the distance in Σ of any two points in $I^+\big(\gamma(-\mu), M_{\text{int}}\big) \cap \Sigma$ is bounded by C_d.

Explicitly, this means that for all $p, q \in I^+\big(\gamma(-\mu), M_{\text{int}}\big) \cap \Sigma$ we have

$$d_\Sigma(p,q) := \inf_{\substack{\gamma:[0,1]\to\Sigma \text{ piecewise smooth} \\ \gamma(0)=p \text{ and } \gamma(1)=q}} \Big\{\int_0^1 \sqrt{\overline{g}(\dot{\gamma}(s), \dot{\gamma}(s))}\, ds\Big\} \leq C_d\,,$$

where $\overline{g}$ is the induced metric on Σ.

Let Σ be a Cauchy hypersurface of M_{int} and assume that $\gamma(-\mu) \in I^-(\Sigma, M_{\text{int}})$ – otherwise there is nothing to show. For all $\underline{x} \in (-\varepsilon_1, \varepsilon_1)^d$ the future directed timelike curves

$$\sigma_{\underline{x}} : \big(-\varepsilon_0, f(\underline{x})\big) \to M_{\text{int}}\,, \qquad \sigma_{\underline{x}}(s) = (\iota^{-1} \circ \tilde{\varphi}^{-1})(s, \underline{x})$$

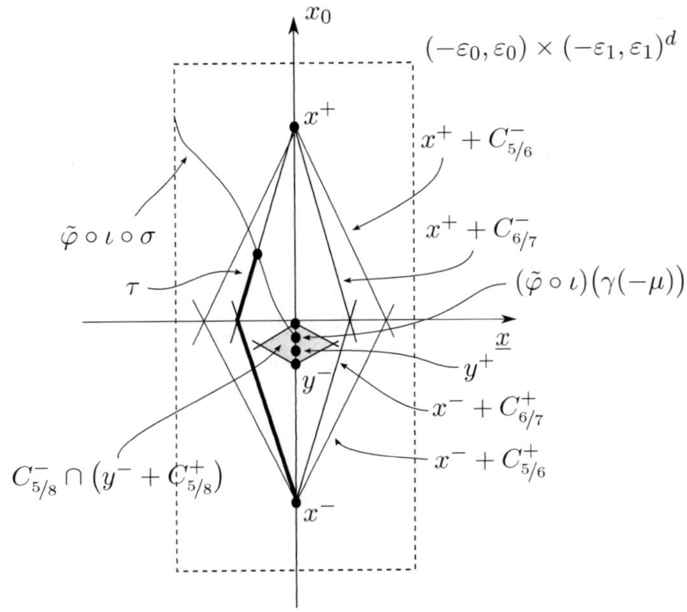

Figure 5. The proof of Step 1 – by contradiction

are future inextendible in M_{int}. Moreover, by the second point of Step 1 they start in $I^-\big(\gamma(-\mu), M_{\text{int}}\big) \subseteq I^-(\Sigma, M_{\text{int}})$. It now follows from Σ being a Cauchy hypersurface of M_{int} that for each $\underline{x} \in (-\varepsilon_1, \varepsilon_1)^d$ the curve $\sigma_{\underline{x}}$ intersect Σ exactly once - say at $s = h(\underline{x})$. This defines a function $h : (-\varepsilon_1, \varepsilon_1)^d \to (-\varepsilon_0, \varepsilon_0)$.

Step 2.1: We show that $h : (-\varepsilon_1, \varepsilon_1)^d \to (-\varepsilon_0, \varepsilon_0)$ is smooth.

Let $\underline{x}_0 \in (-\varepsilon_1, \varepsilon_1)^d$. Since $\tilde{\varphi}\big(\iota(\Sigma) \cap \tilde{U}\big)$ is a smooth submanifold of $R_{\varepsilon_0,\varepsilon_1}$, there exists a smooth submersion $u : W \to \mathbb{R}$, where $W \subseteq R_{\varepsilon_0,\varepsilon_1}$ is an open neighbourhood of $\big(h(\underline{x}_0), \underline{x}_0\big)$, such that $\tilde{\varphi}\big(\iota(\Sigma) \cap \tilde{U}\big) \cap W = \{u = 0\}$. Since Σ is a Cauchy hypersurface, the timelike vector field ∂_0 can be nowhere tangent to $\tilde{\varphi}\big(\iota(\Sigma) \cap \tilde{U}\big)$. It thus follows that $\partial_0 u|_{\big(h(\underline{x}_0), \underline{x}_0\big)} \neq 0$. By the implicit function theorem, there is now a smooth function $v : V \to \mathbb{R}$, where $V \subseteq (-\varepsilon_1, \varepsilon_1)^d$ is an open neighbourhood of $\underline{x}_0$, such that $u\big(v(\underline{x}), \underline{x}\big) = 0$. Thus, we must have $h|_V = v$ – and hence, h is smooth.

Step 2.2: We show that there exists a $0 < C_{\text{slope}} < \infty$ such that $|\partial_i h(\underline{x})| \leq C_{\text{slope}}$ holds for all $\underline{x} \in (-\varepsilon_1, \varepsilon_1)^d$ and for all $i = 1, \ldots, d$.

Since the tangent space of a Cauchy hypersurface does not contain timelike vectors, we obtain the inequality

$$0 \leq \tilde{g}\Big((\partial_i h)\partial_0 + \partial_i, (\partial_i h)\partial_0 + \partial_i\Big) = (\partial_i h)^2 \tilde{g}_{00} + 2(\partial_i h)\tilde{g}_{0i} + \tilde{g}_{ii} \tag{4.16}$$

for all $i = 1, \ldots, d$. Equality in (4.16) holds for

$$(\partial_i h)_\pm = \frac{-\tilde{g}_{0i} \mp \sqrt{(\tilde{g}_{0i})^2 - \tilde{g}_{ii}\tilde{g}_{00}}}{\tilde{g}_{00}} .$$

Recall that $\tilde{g}_{00} < -1 + \delta_0 < 0$ and $|\tilde{g}_{\mu\nu}| \leq 1 + \delta_0$. Hence, we obtain the uniform bound

$$\max\Big\{|(\partial_i h)_+|, |(\partial_i h)_-|\Big\} \leq C_{\text{slope}} ,$$

where $0 < C_{\text{slope}} < \infty$ is a constant depending on δ_0 (but not on h). Moreover, together with $\tilde{g}_{00} < 0$, the inequality (4.16) implies

$$(\partial_i h)_- \leq (\partial_i h) \leq (\partial_i h)_+$$

and thus $|\partial_i h| \leq C_{\text{slope}}$ for all $i = 1, \ldots, d$.

We now define $\omega : (-\varepsilon_1, \varepsilon_1)^d \to R_{\varepsilon_0,\varepsilon_1}$ by $\omega(\underline{x}) = \big(h(\underline{x}), \underline{x}\big)$. Clearly, ω parametrises a smooth submanifold $\tilde{S}$ of $R_{\varepsilon_0,\varepsilon_1}$ which, via $\iota^{-1} \circ \tilde{\varphi}^{-1}$, is isometric to an open subset S of $\Sigma \subseteq M_{\text{int}}$. The ambient metric $\tilde{g}$ on $R_{\varepsilon_0,\varepsilon_1}$ induces a metric $\overline{g}$ on $\tilde{S}$. Note that this metric is not necessarily positive definite. Its components with respect to the chart ω^{-1} are denoted by $\overline{g}_{ij}$, where $i, j \in \{1, \ldots, d\}$.

Step 2.3: We show that there exists a constant $0 < C_{\overline{g}} < \infty$ such that the uniform bound $|\overline{g}_{ij}(\underline{x})| \leq C_{\overline{g}}$ holds for all $\underline{x} \in (-\varepsilon_1, \varepsilon_1)^d$ and for all $i, j = 1, \ldots, d$.

We compute

$$\overline{g}_{ij} = \big(\omega^* \tilde{g}\big)_{ij} = \tilde{g}_{\mu\nu} \frac{\partial \omega^\mu}{\partial x_i} \frac{\partial \omega^\nu}{\partial x_j} = \tilde{g}_{00} \frac{\partial h}{\partial x_i} \frac{\partial h}{\partial x_j} + \tilde{g}_{0j} \frac{\partial h}{\partial x_i} + + \tilde{g}_{i0} \frac{\partial h}{\partial x_j} + \tilde{g}_{ij} \,.$$

It now follows from $|\tilde{g}_{\mu\nu}| \leq 1 + \delta_0$ and Step 2.2 that there exists a constant $0 < C_{\overline{g}} < \infty$ such that $|\overline{g}_{ij}| \leq C_{\overline{g}}$ holds for all $i, j = 1, \ldots, d$.

We can now finish Step 2. Let $p, q \in I^+\big(\gamma(-\mu), M_{\text{int}}\big) \cap \Sigma$ be given. By property 1 of Step 1 there exists $\underline{x}, \underline{y} \in (-\varepsilon_1, \varepsilon_1)^d$ with $\omega(\underline{x}) = \tilde{\varphi}\big(\iota(p)\big)$ and $\omega(\underline{y}) = \tilde{\varphi}\big(\iota(q)\big)$. We connect the points $\underline{x}$ and $\underline{y}$ by a straight line in $(-\varepsilon_1, \varepsilon_1)^d$, i.e., by $\sigma : [0, 1] \to (-\varepsilon_1, \varepsilon_1)^d$, where $\sigma(s) = \underline{x} + s(\underline{y} - \underline{x})$. It follows that

$$\begin{aligned}
L(\sigma) &= \int_0^1 \sqrt{\overline{g}\big(\dot{\sigma}(s), \dot{\sigma}(s)\big)}\, ds \\
&= \int_0^1 \sqrt{\Big(\sum_{i,j=1}^d (\underline{y}_i - \underline{x}_i) \overline{g}_{ij}\big(\sigma(s)\big)(\underline{y}_j - \underline{x}_j)\Big)}\, ds \\
&\leq \int_0^1 \sqrt{\Big(\sum_{i,j=1}^d \sqrt{d}\varepsilon_1 \cdot C_{\overline{g}} \cdot \sqrt{d}\varepsilon_1\Big)}\, ds \\
&= \varepsilon_1 d^{3/2} \sqrt{C_{\overline{g}}} \,.
\end{aligned}$$

Note that this bound does not depend on the points $\underline{x}, \underline{y} \in (-\varepsilon_1, \varepsilon_1)^d$. It follows that $\iota^{-1} \circ \tilde{\varphi}^{-1} \circ \omega \circ \sigma$ is a smooth curve in Σ that connects p and q and the length of which is less or equal to $\varepsilon_1 d^{3/2} \sqrt{C_{\overline{g}}}$. Since

$$\begin{aligned}
d_\Sigma(p, q) &= \inf_{\substack{\gamma:[0,1]\to\Sigma \text{ piecewise smooth} \\ \gamma(0)=p \text{ and } \gamma(1)=q}} \Big\{ \int_0^1 \sqrt{\overline{g}\big(\dot{\gamma}(s), \dot{\gamma}(s)\big)}\, ds \Big\} \\
&\leq \inf_{\substack{\gamma:[0,1]\to S \text{ piecewise smooth} \\ \gamma(0)=p \text{ and } \gamma(1)=q}} \Big\{ \int_0^1 \sqrt{\overline{g}\big(\dot{\gamma}(s), \dot{\gamma}(s)\big)}\, ds \Big\} \\
&\leq \varepsilon_1 d^{3/2} \sqrt{C_{\overline{g}}} \,,
\end{aligned}$$

IOP Conf. Series: Journal of Physics: Conf. Series **968** (2018) 012012 doi:10.1088/1742-6596/968/1/012012

this concludes Step 2 with $C_d = \varepsilon_1 d^{3/2}\sqrt{C_{\overline{g}}}$.

Step 3: We show that the result of Step 2 is in contradiction with the geometry of $(M_{\text{int}}, g_{\text{int}})$.

Let ω_0 be the projection of $\gamma(-\frac{\mu}{2})$ to $\mathbb{S}^{d-1}$, $t_0 := t\big(\gamma(-\frac{\mu}{2})\big)$, and $r_0 := r\big(\gamma(-\frac{\mu}{2})\big)$. By the openness of $I^+\big(\gamma(-\mu), M_{\text{int}}\big)$ there exists a $\lambda > 0$ such that $(t_0 - \lambda, t_0 + \lambda) \times r_0 \times \omega_0 \subseteq I^+\big(\gamma(-\mu), M_{\text{int}}\big)$. Since $-\partial_r$ is future directed timelike, we also have

$$[t_0 - \lambda, t_0 + \lambda] \times (0, r_0) \times \omega_0 \subseteq I^+\big(\gamma(-\mu), M_{\text{int}}\big) \,.$$

Next, we consider the family of Cauchy hypersurfaces $\Sigma_n := \{r = \frac{1}{n}\}$, where $n \in \mathbb{N}_{>n_0}$ for some large $n_0 \in \mathbb{N}$, together with the family of pairs of points

$$p_n := (t_0 - \lambda, \frac{1}{n}, \omega_0) \in I^+\big(\gamma(-\mu), M_{\text{int}}\big) \qquad \text{and} \qquad q_n := (t_0 + \lambda, \frac{1}{n}, \omega_0) \in I^+\big(\gamma(-\mu), M_{\text{int}}\big) \,.$$

The induced metric $\overline{g}_n$ on Σ_n is given by

$$\overline{g}_n = -\big(1 - 2m \cdot n^{(d-2)}\big)\, dt^2 + n^{-2}\, \mathring{\gamma}_{d-1} \,.$$

It is easy to see that the shortest curve connecting p_n with q_n in Σ_n is given by $\gamma_n : (-\lambda, \lambda) \to \Sigma_n$,

$$\gamma_n(s) = (t_0 + s, \frac{1}{n}, \omega_0) \,.$$

The length $L(\gamma_n)$ of γ_n is given by

$$\begin{aligned} L(\gamma_n) &= \int_{-\lambda}^{\lambda} \sqrt{\overline{g}_n\big(\dot{\gamma}_n(s), \dot{\gamma}_n(s)\big)}\, ds \\ &= \int_{-\lambda}^{\lambda} \sqrt{2m \cdot n^{(d-2)} - 1}\, ds \\ &= 2\lambda\sqrt{2m \cdot n^{(d-2)} - 1} \,. \end{aligned}$$

Since $d \geq 3$, it thus follows that

$$d_{\Sigma_n}(p_n, q_n) = 2\lambda\sqrt{2m \cdot n^{(d-2)} - 1} \to \infty \qquad \text{for } n \to \infty \,,$$

in contradiction to Step 2. This concludes the proof of Theorem 1.

References

[1] Sbierski J 2015 *arXiv:1507.00601, to appear in the Journal of Differential Geometry*
[2] Galloway G, Ling E and Sbierski J 2017 *Comm. Math. Phys. https://doi.org/10.1007/s00220-017-3019-2*
[3] Galloway G and Ling E 2017 *Ann. Henri Poincaré* **18** 3427–3477
[4] O'Neill B 1983 *Semi-Riemannian geometry* (Academic Press Inc)